Katja Jöchen

Homogenization of the Linear and Non-linear Mechanical Behavior of Polycrystals

Schriftenreihe
Kontinuumsmechanik im Maschinenbau
Band 4

Karlsruher Institut für Technologie (KIT)
Institut für Technische Mechanik
Bereich Kontinuumsmechanik

Hrsg. Prof. Dr.-Ing.habil. Thomas Böhlke

Eine Übersicht über alle bisher in dieser Schriftenreihe erschienenen Bände finden Sie am Ende des Buchs.

Homogenization of the Linear and Non-linear Mechanical Behavior of Polycrystals

by
Katja Jöchen

Dissertation, Karlsruher Institut für Technologie (KIT)
Fakultät für Maschinenbau
Tag der mündlichen Prüfung: 23. November 2012

Impressum

Karlsruher Institut für Technologie (KIT)
KIT Scientific Publishing
Straße am Forum 2
D-76131 Karlsruhe
www.ksp.kit.edu

KIT – Universität des Landes Baden-Württemberg und
nationales Forschungszentrum in der Helmholtz-Gemeinschaft

KIT Scientific Publishing 2013
Print on Demand

ISSN 2192-693X
ISBN 978-3-86644-971-8

Homogenization of the Linear and Non-linear Mechanical Behavior of Polycrystals

Zur Erlangung des akademischen Grades

Doktor der Ingenieurwissenschaften

der Fakultät für Maschinenbau
Karlsruher Institut für Technologie (KIT)

genehmigte

Dissertation

von

Dipl.-Ing. Katja Jöchen

Tag der mündlichen Prüfung: 23. November 2012
Hauptreferent: Prof. Dr.-Ing. Thomas Böhlke
Korreferent: Prof. Dr. Laurent Delannay

Zusammenfassung

Das mechanischen Verhalten mikrostrukturierter Materialien wird maßgeblich durch das konstitutive Verhalten und die Anordnung der Einzelphasen bestimmt. Ohne die Berücksichtigung von mikrostrukturellen Parametern lässt sich das mechanische Verhalten solcher Materialien in den meisten Fällen nur unzulänglich vorhersagen.

Diese Arbeit befasst sich im Wesentlichen mit der numerisch effektiven Simulation der Materialantwort polykristalliner Aggregate. Dafür wird in Verbindung mit der Kristallplastizität ein neues nichtlineares Homogenisierungsschema genutzt, welches auf stückweise konstanten Spannungspolarisationen bezüglich eines homogenen Vergleichsmediums basiert und einer Verallgemeinerung des Hashin-Shtrikman Ansatzes entspricht. In diesem Mean-Field-Ansatz wird die Ein- und Zweipunktstatistik der Mikrostruktur berücksichtigt, welches einen Kompromiss zwischen der rechenintensiven ortsaufgelösten Analyse von Mikrostrukturen und der Anwendung einfacher Mischungstheorien darstellt.

In mehreren geometrisch und physikalisch linearen und nichtlinearen Anwendungsfällen werden verschiedenene Homogenisierungsstrategien getestet und bewertet. Insbesondere wird die Vorhersage der Texturentwicklung in Polykristallen untersucht sowie erste Anwendungen stark gekoppelter zweiskaliger Simulationen gezeigt, in denen das Homogenisierungsschema am Integrationspunkt eines FE-Modells zur Anwendung kommt. Im letzteren Fall wird die experimentell gemessene Ausgangstextur mit einem neuen Reduktionsverfahren auf die wichtigsten Texturkomponenten reduziert, um die numerische Effizienz der Zweiskalensimulation zu gewährleisten.

Summary

The mechanical behavior of materials with microstructure is significantly governed by the constitutive behavior of the single constituents as well as their arrangement. Without consideration of microstructural parameters, usually the mechanical response of those materials is predicted inadequately.

This work is mainly dedicated to the numerically efficient simulation of the material response of polycrystalline aggregates. Therefore, crystal plasticity is combined with a new non-linear homogenization scheme, which is based on piecewise constant stress polarizations with respect to a homogeneous reference medium and corresponds to a generalization of the Hashin-Shtrikman scheme. This mean field approach accounts for the one- and two-point statistics of the microstructure and, hence, complies with a compromise between the computationally costly spatially resolved analysis of the microstructure and the application of elementary mixture theories.

Several geometrically and physically linear and non-linear examples are considered by applying different homogenization schemes and evaluating the results. Especially, the prediction of texture evolution in polycrystals is investigated and first applications of strongly coupled two-scale simulations are shown in which the homogenization scheme is incorporated at the integration points of a finite element model. In the latter example, experimentally measured texture data is representatively reduced to major texture components by a new reduction technique in order to ensure the numerical efficiency of the two-scale simulations.

Acknowledgments

I would like to express my sincere gratitude to my supervisor Prof. Dr.-Ing. Thomas Böhlke for sparking my interest in micromechanics, the confidence, guidance and unreserved encouragement.

Special thanks go to Prof. Dr. Laurent Delannay, who kindly agreed to be the co-reviewer of my thesis, provided valuable experimental data and offered the opportunity for a short research stay at UCL.

Thank you, Prof. Dr.-Ing. Oliver Kraft, for being the chair of the committee and for the convenient meeting prior to the defense.

Many thanks to my colleagues for the supportive atmosphere, the distractive coffee breaks and the enjoyable company during lunch time.

Especially, I want to thank my room mate Barthel for the cheerful sentiment in our office. I will never forget our regular high-level talks on orientations, sometimes feeling a bit unoriented.

Ute and Helga, I warmly thank you for your help in administrative things, for proof reading and your general assistance in all situations.

My heartfelt thanks go to my parents, my grand-parents and my sister, who absolutely supported me regardless of which idea crossed my mind.

Most of all, I am indebted to Dirk for his everlasting encouragement and to Hanna for the necessary diversion. You are the world to me.

Contents

7 Geometrically Non-linear Applications of the Non-linear Hashin-Shtrikman-type Scheme   121

8 Summary and Conclusions   151

A Appendix   155

Bibliography   171

Chapter 1

Preliminaries

1.1 Motivation

"The behavior of metals in the inelastic states cannot be explained by theories which do not assume a microstructure." Hencky (1932)

Actually, this quotation very succinctly emphasizes the motivation and challenges for modeling the constitutive response of materials with microstructure. To expatiate on the incentive of this thesis, the above quotation should be extended such that not only the inelastic states, but even the accurate description of the elastic behavior of materials (not restricted to metals) requires the incorporation of microstructural information.

Strictly speaking, with the exception of perfect single crystals, heterogeneities can be found in all materials at any scale. Although this fact naturally results in difficulties for properly describing these materials by simple models, it offers also a tremendous potential for designing materials with enhanced material behavior.

Fig. 1.1 depicts some selected microstructures of materials for industrial applications ranging from fiber and particle reinforced materials to multi-phase crystalline aggregates. When aiming to describe the overall constitutive behavior of such materials, it is unconditionally necessary to not only account for the constitutive behavior of the individual phases but also for their arrangement and shapes. Thinking of, e.g., uni-directional fiber reinforcements, their special arrangement is solely generated in order to obtain a macroscopic anisotropy even though the fibers and matrix materials may be described by isotropic constitutive laws. On the other hand, an anisotropy on the macroscale can also be an outcome of manufacturing processes as, e.g., in rolled sheets where initially macroscopic isotropic crystalline aggregates are severely deformed re-

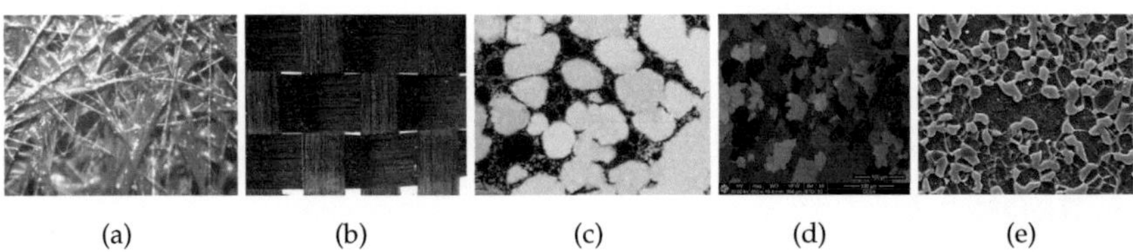

(a) (b) (c) (d) (e)

Figure 1.1: Examples of microstructures: (a) LFT - Long fiber reinforced thermoplastics (ITM, KIT), (b) fabric for RTM - Resin Transfer Molding (ITM, KIT), (c) Metal matrix composite (Carrasco et al., 2011), (d) Ferritic steel DC04 (IAM-WBM, KIT), (e) Dual-phase steel (Delincé et al., 2007)

sulting in crystallographic and morphological textures macroscopically causing deformation-induced anisotropy.

The mathematical description of the behavior of microheterogeneous materials on the macroscale requires to properly account for the microstructural phenomena. The first step is to accurately transfer the influence of the macroscopic onto the microscopic fields being called localization. In the second step, the constitutive response of the phases of the microstructure needs to be computed and averaged such as to transfer the results to the macroscale, the latter being called homogenization. Several approaches exist for describing the localization and homogenization steps, the most accurate and realistic certainly being full-field simulations where the microstructure is completely resolved (apart from the necessity of discretization) and phase interactions are accounted for. The main drawback of this approach is the enormous computational effort which is still not compatible with industrial needs even though the computational power is developing rapidly.

Considerably less computational resources are needed when applying mean-field approaches. Mainly being based on the Eshelby solution (Eshelby, 1957), proving that the mechanical fields in a single ellipsoidal inclusion/inhomogeneity being placed in an infinitely extended matrix are homogeneous in the elastic case, phase averages of stresses and strains are related by these approaches drastically reducing the number of degrees of freedom. In this case, bounds and estimates for the effective (macroscopic) material response can be derived, the quality of which strongly depends on the assumptions made.

This thesis is dedicated to mathematically describing the elastic and inelastic response of single-phase crystalline aggregates by using mean-field approaches leading to the development of numerically efficient models that can be applied at the integration points of finite elements. Therefore, Chapter 1 provides the motivation for this work, the state of the art and the notation. In Chapter 2, the theoretical basics are thoroughly introduced ranging from the fundamentals of continuum mechanics over representations of crystal orientations to the homogenization theory. Real and synthetically generated microstructures are described in Chapter 3 with special emphasis to Voronoi-type mosaics. In Chapter 4, two approaches for reducing large orientation data sets measured by, e.g., EBSD or X-ray diffraction are introduced. Chapter 5 is dedicated to homogenization methods for describing the physically linear and Chapter 6 the physically non-linear material response of crystalline aggregates. In both chapters, different approaches are presented and applied in the infinitesimal strain regime. Additionally, representative for applications in the finite strain regime, Chapter 7 highlights the texture prediction by using the proposed non-linear Hashin-Shtrikman method and gives first examples for strongly coupled two-scale simulations. Chapter 8 provides the summary of the thesis.

1.2 State of the art

This section aims to give an overview, certainly far from being complete, on established and approved theories as well as recent publications being related to the topics addressed in this thesis. In the sequel, works on the generation of artificial microstructures, the reduction of large measured orientation data sets, homogenization approaches, models for predicting texture evolution and the extension to two-scale simulations are briefly discussed.

Synthetic microstructures

The mechanical response of microstructured materials strongly depends on the size, shape and orientation distributions of the phases so that the microstructure should be replicated by the computational model as precise as possible. Due to the enormous effort of measuring and digitizing real three-dimensional microstructures (e.g., destructively by focused ion beam sectioning (FIB)), it is required to generate artificial microstructures reproducing as accurately as possible the important microstructural characteristics.

In order to model polycrystalline structures, Voronoi-type tessellations (Aurenhammer, 1991) are used very frequently, although the associated highly idealized grain-growth process is rarely met in reality (Ohser and Mücklich, 2000). While the pure Poisson-Voronoi tessellation results in non-natural acute angles of the polyhedral grains, some modifications are proposed in the literature, to improve the Voronoi mosaics. On the one hand, Laguerre-type tessellations (e.g., Lautensack and Zuyev (2008)) are proposed to model, e.g., microstructures of severely deformed materials (Jafari and Kazeminezhad, 2011). Zhang et al. (2012) propose to apply a tessellation method with regularity control and, therefore, also grain size distribution control in order to vary between the Poisson-Voronoi mosaic and a regular hexagonal tessellation. For computational efficiency, e.g., Fritzen et al. (2009) as well as Quey et al. (2011) propose improved meshing strategies for Voronoi-type microstructures. Applications with additional morphological aspects being present in rolled sheets and materials with grain size gradients are treated in their papers. Furthermore, an adapted FEM technique is introduced by Ghosh and Moorthy (1995), which is the VCFEM (Voronoi cell finite element method), using the Voronoi cells as polygonal finite elements, e.g., in electromechanical simulations (Jayabal and Menzel, 2012).

All the aforementioned tessellations reproduce the real microstructures only to a certain extent, so that, nevertheless, some works are dedicated to including experimental information into the microstructural models. While Bhandari et al. (2007) uses the complete FIB information of three-dimensional microstructures being reconstructed by NURBS to form smooth grain boundaries and allowing for coarsening the FE mesh, other groups attempt to incorporate statistical information of the microstructure. Saylor et al. (2004) uses images of two orthogonal sections of the investigated real microstructures so as to gain information on the distribution of grain shapes being idealized by ellipsoids. Based on this distribution, three-dimensional synthetic microstructures are generated which are statistically representative of the real materials not only in terms of volume fraction but also of shape distribution. The ansatz of Groeber et al. (2008) is much more elaborate, since it uses the experimental distributions of higher-order moments. As input for their scheme, besides the aforementioned distributions, also correlation information for, e.g., aspect ratio to grain volume or number of neighbors to grain volume are used, which initially necessitates a very broad analysis of the investigated material.

Reduction of orientation data

With regard to the precise but efficient computation of polycrystalline material response, it is required to incorporate a representative but small orientation data set of the experimental microstructure into the model. Several methods are recorded in literature, so as to identify a small set of texture components from a measured orientation data set or orientation distribution function.

Tóth and Van Houtte (1992) proposed to divide the orientation space in a regular grid and to select grid points by a probabilistic approach based on the measured orientation distribution in order to generate a set of prescribed number of discrete orientations with equal volume fractions. Improved texture approximations are obtained by the extended approach of Melchior and Delannay (2006), who used the latter method but, furthermore, merged discrete orientations of similar orientation ending up in a smaller data set of components with varying volume fractions.

In the method of Cho et al. (2004) the texture approximation is based on a fixed set of typical components of cubic metal textures (e.g., brass, copper, cube components). In their approach, the texture components are described by von Mises-Fisher distributions with a fixed half-width, and the associated volume fractions are determined as a sum of the intensities of the measured orientation distribution function within a certain misorientation range. Also Raabe and Roters (2004) utilize ideal texture components of metal textures at the integration points of finite element models. In a second step, they slightly rotate these ideal orientations so that finally in the whole model (e.g, in the rolled sheet), the measured orientation distribution is reproduced. To overcome the problem of too sharp textures, a so-called random background texture is additionally assigned to the model.

Böhlke et al. (2006a) used a mixed integer quadratic programming procedure to identify main texture components with appropriate volume fractions. Once more, an additional fraction of isotropic texture is introduced to weaken the approximate CODF. Due to the applied optimization procedure, a very small number of texture components is identified to be sufficient to approximate the experimental texture.

The hybrid approach of Eisenlohr and Roters (2008) (see also Roters et al. (2010)) combines a deterministic and probabilistic scheme in order to sample a given number of orientations with equal weights from a discrete orientation distribution function. Only this combined procedure allows to reconstruct the measured texture without significant sharpening for a small number of texture components.

Recently, Jöchen and Böhlke (2011) proposed to apply a sectioning technique of the orientation space with an appropriate averaging scheme so as to directly extract a small representative texture data set from large measured data sets, e.g., obtained by EBSD. This method, a modification, and an additional clustering technique are intensively discussed in Chapter 4.

Homogenization techniques

The development of homogenization schemes to predict the material response of microheterogeneous materials is a topic of research since more than 120 years and, correspondingly, lead and leads to large numbers of investigated approaches. The main works in the context of linear materials, as the ones by Voigt (1889); Reuss (1929); Eshelby (1957); Hashin and Shtrikman (1962), building the basis of many recent approaches, are explicitly reflected in Chapter 5. For extensive overviews on predominantly linear homogenization schemes, the reader is referred to works like Willis (1981); Mura (1987); Nemat-Nasser and Hori (1993); Torquato (2002); Kanouté et al. (2009).

Since the characterization of the physically linear behavior of microheterogeneous materials is fairly well developed based on the exact validity of the Eshelby solution (see also Chapter 5), in the sequel, the very recent approaches on mainly non-linear homogenization as well as texture prediction by homogenization schemes are reviewed.

The non-linear extension of the promising linear homogenization methods is the standard way to compute physically non-linear microheterogeneous material behavior. Therefore, e.g., the self-consistent method in combination with a local linearization is applied being usually classified into incremental (Hill, 1965a; Hutchinson, 1976), tangent (Molinari et al., 1987; Lebensohn and Tomé,

1993; Mercier and Molinari, 2009), secant (Berveiller and Zaoui, 1979; Qiu and Weng, 1992; Suquet, 1995), affine (Masson et al., 2000; Brenner et al., 2001) and variational (Ponte Castañeda and Nebozhyn, 1997; Nebozhyn et al., 2001) approaches, which are, e.g., compared by Liu et al. (2005) for the case of texture evolution in halite. The problems of incremental and secant approaches yielding too stiff results and probably even violating bounds (Gilormini, 1996) have been overcome by taking into account second-order moments of the fields (Suquet, 1995), which lead to an equivalent representation of the effective behavior as derived by the variational procedure of Ponte Castañeda (1991).

Another classification of non-linear homogenization procedures is based on the usage of homogeneous or heterogeneous comparison materials, the first being often referred to as HEM (homogeneous equivalent medium) and the latter as LCC (linear comparison composite). While the classical extensions of the linear methods use (non)-linear homogeneous comparison media (e.g., Talbot and Willis (1985); Lebensohn and Tomé (1993)), a newer class of variational methods is dedicated to optimally identify a LCC so as to incorporate heterogeneities of inelastic fields in a piecewise constant sense and subsequently apply a linear homogenization method (Ponte Castañeda, 1991; Ponte Castañeda and Suquet, 1998). For higher accuracy, the LCC approach is enriched by second-order moments to obtain results being exact to second-order in the contrast (Ponte Castañeda, 2002; Idiart and Ponte Castañeda, 2007).

Contrary to the previously discussed semi-analytical methods, partly and purely numerical approaches make use of full-field solutions of spatially resolved microstructures. The 'Transformation Field Analysis' (TFA, Dvorak and Benveniste (1992)) partitions the microstructure in several subdomains and assumes piecewise constant inelastic strains which are updated based on phase averages of stresses and strains. The elastic localization tensors known from elasticity are used to determine the local fields and need, therefore, only be computed once for all (usually done by FEM). For obtaining an acceptable solution, the number of subdomains needs to be significantly larger than the number of phases. This fact gave cause to introducing non-uniform inelastic fields being a finite set of inelastic modes linearly combined to approximate the inelastic strain heterogeneity. This method is referred to as NTFA (Michel and Suquet, 2003; Fritzen and Böhlke, 2010). In opposition to the aforementioned semi-analytical schemes, these methods are computationally more expensive since they require some elastic(-plastic) full-field finite element solutions of a fixed microstructure in advance to the solution of the present boundary value problem. Nevertheless, they give very precise and reliable solutions and are still much more efficient than total full-field RVE computations when different load cases are to be considered. To round out the remarks on numerical methods, it is worth mentioning the image-processing inspired FFT (Fast Fourier Transform) techniques (e.g., Moulinec and Suquet (1998); Lebensohn (2001)), which also provide full-field solutions being computationally much more efficient than crystal plasticity FEM (Prakash and Lebensohn, 2009; Roters et al., 2010).

Prediction of texture evolution

A broad and important field of application of homogenization methods represents the prediction of texture evolution during the deformation of polycrystalline aggregates. Since a representative section of the microstructure with numerous details and, therefore, numerous degrees of freedom as well as large deformations need to be considered, in spite of increasing computational power, full-field computations are pretty costly especially with regard to two-scale applications. Fast but sufficiently accurate methods are required to fulfill these needs. Nevertheless, finite element based texture prediction (e.g., Anand, 2004; Sundararaghavan and Zabaras, 2006) plays an important role, not at least for assessing the validity of applying homogenization models (e.g., Liu et al., 2003).

As already noted about 40 years ago, the rolling textures of fcc crystal aggregates are characterized by a pure metal or alloy-type texture, also known as copper- and brass- type textures (Leffers, 1968; Bunge and Tobisch, 1972; Hirsch and Lücke, 1988a; Wierzbanowski et al., 1992). On the other hand, the simple homogenization approaches of Taylor (1938) and Sachs (1928) are known to predominantly pronounce the copper or the brass component, respectively. Therefore, relaxing the strong constraints of these methods raised hope to computationally obtain textures in between these extremal cases (e.g., Kocks and Chandra, 1982; Hirsch and Lücke, 1988b; Leffers, 1995; Van Houtte et al., 1999).

With regard to this aspect, Delannay et al. (2002) worked on the texture prediction in cold rolled aluminum by using the full- and relaxed-constraint Taylor theories and the LAMEL model (Van Houtte et al., 1999) and compared the results to FEM results and experiments. They found that grains initially oriented near the cube component generally rotate too rapid to the copper component, and usually the brass component is insufficiently reproduced compared to experiments. The relaxed Taylor methods do not satisfactorily reproduce the pronounced locations of the CODF, especially the β-fiber. The brass component is found to become more pronounced if the RD-TD shear is relaxed.

Sarma and Dawson (1996) propose a model in order to non-uniformly distribute the deformation and the spin among the crystals of the polycrystalline aggregate. Based on the compliances of the crystal and of its neighborhood, the specific partitioning rule is constructed, wherein the deviation of the deformation rate from the macroscopic one is linked to the law for determining the lattice spin of the crystal. The influence of the number of neighboring grains accounted for in the compliance computation is discussed in detail. Furthermore, texture prediction in case of rolling and simple shear is compared to results from the Taylor approach and finite element computations as well as experiments. The neighborhood compliance model shows to predict more diffuse textures than the Taylor method, leading to a more realistic pronunciation of texture components when comparing to FEM results and experiments.

Several research groups work on using different variants of the self-consistent method for the prediction of texture evolution. Lebensohn and Leffers (1999) deal with different types of lattice rotation to obtain textures during rolling. They use two rules of lattice rotation from Hosford (1977), so as to have the so-called mathematical analysis (MA) and a plane-strain analysis (PSA), the first being appropriate for equiaxed grains and the latter for elongated flat grains. Additionally, they use the VPSC scheme (Lebensohn and Tomé, 1993) with updating of grain shapes and obtain an intermediate texture prediction between the copper dominated MA and the brass-dominated PSA. They believe that probably materials yielding brass-type textures show updating grain shapes during deformation while grains in materials with copper-type textures divide, so effectively they remain equiaxed.

Lipinski and Berveiller (1989) propose to use a tangent modulus which is not possessing the typical symmetries so as to compute an asymmetric integral operator for appropriately modeling lattice rotations. In their model, this non-symmetric elasto-plastic modulus relates nominal stress rate and the velocity gradient. Therefore, since by using the velocity gradient, also the asymmetric spin part therein is coupled to the stress rate conflicting with the definition of a constitutive law. Neil et al. (2010) use an elasto-plastic SC scheme to account for large strain kinematics in an approximate way. They model texture evolution for copper and a stainless steel and find that their scheme is less rigorous but computationally more efficient than the one by Lipinski and Berveiller (1989). Examining a fcc duplex steel, the copper to brass texture transition capabilities of the SC model have been improved by incorporating micro shear banding (Jia et al., 2011).

The tangent VPSC scheme introduced by Molinari et al. (1987) and used with an anisotropic linear comparison material by Lebensohn and Tomé (1993, 1994) is found to underestimate the effective flow stress in highly non-linear anisotropic polycrystals (Lebensohn and Tomé, 1993; Nebozhyn et al., 2001). Nevertheless, the method was successfully applied to predict titanium deformation textures, but for halite, the result is similar to the Taylor prediction (Liu et al., 2003). For textures in halite, the second-order SC scheme (Ponte Castañeda, 2002; Liu and Castañeda, 2004a,b) was successfully applied by Liu et al. (2005). Also Wenk (1999) worked on the material behavior of geologic low-crystal symmetry materials (e.g., ice, halite, calcite, quartz, olivine) and showed that the VPSC scheme yields pretty improved results compared to the Taylor prediction when comparing with experiments. Lebensohn et al. (2011) compared the capabilities of VPSC (tangent, affine and second-order) for texture prediction, where it has been shown that the second-order approach is in best agreement with FFT solutions. Furthermore, very sophisticated processes and materials are investigated using self-consistent methods, as, e.g., the ECAE test (equal channel angular extrusion) by Signorelli et al. (2006) or texture prediction of semi-crystalline polymers (Nikolov et al., 2006). Also other intermediate models, as the ϕ-model developed by Ahzi and M'Guil (2008), have been compared to the SC approach by M'Guil et al. (2009) and used for the prediction of fcc rolling textures and lankford coefficients by M'Guil et al. (2010) and of bcc textures by M'Guil et al. (2011).

Two-scale simulations

A strong need exists for incorporating microstructural information in simulations of structural components especially in forming operations with regard to reliable predictions of the material response (overview given in Geers et al.,

2010, also on second-order two-scale simulations including gradients of the deformation gradient being not discussed here). The most accurate approach in this field is the usage of the FE^2 method (Renard and Marmonier, 1987; Miehe et al., 1999; Feyel, 2003), in which a spatially resolved RVE of the microstructure discretized by finite elements is linked to the integration points of finite elements of macroscopic parts. Thus, the boundary conditions of the micro FE model are prescribed by the local conditions of the macro FE model. Although a pretty authentic behavior is predicted by this approach, the computational times are unaffordable (Van Houtte et al., 2012).

Nevertheless, this approach is subject of current primarily academic research and is applied in a broad field, among which, e.g., the simulation of sheet metal forming limits using the 'Limiting Dome Height' test (Nakamachi et al., 2007; Kuramae et al., 2010) or in civil engineering applications simulating the influence of frost in cement paste (Hain and Wriggers, 2008) can be given as examples. Due to the enormous computational effort needed for solving such two-scale problems, some attempts are made to reduce the costs. For example, Balzani et al. (2010) propose to generate statistically similar RVEs with less complexity in order to reduce the number of degrees of freedom on the microscale or Novak et al. (2012) uses an enhanced FE method for modeling the RVE enabling a coarser mesh even though accounting for inhomogeneities.

A more efficient approach of accounting for the microstructure on the second scale is the application of mean-field theories. The most simple approach is the usage of the assumption of homogeneous deformation of all phases by the Taylor scheme at the integration points of the structural FE model. While Mathur et al. (1990); Aretz et al. (2000) simulated the rolling process of sheet metals, Mathur and Dawson (1989); Kalidindi et al. (1992) used the Taylor-FEM coupling to predict texture evolution during non-uniform bulk-forming, whereas the latter model has been furthermore extended to capture twinning effects (Kalidindi, 1998). Phan Van et al. (2012) simulated deep drawing of a ferritic steel sheet by, contrary to the aforementioned works, coupling the Taylor scheme with an implicit finite element code. Embedding the self-consistent (SC) method in the FE code yields improved results compared to applying the Taylor method, however, with the drawback of increasing computational time. Tomé et al. (2001) and Segurado et al. (2012) simulated bending of a Zr beam by coupling SC to an explicit and implicit FE code, respectively. Also rolling of an Al alloy (Segurado et al., 2012) and deep-drawing of Mg sheets (Walde and Riedel, 2007) have been examined by this approach.

For two-phase-composite applications, also the Mori-Tanaka approach can be used to account for local microstructure in structural FE simulations. Besides the consideration of a smaller number of single phases, the multi-scale investigations of the deformation behavior of, e.g., metal matrix composites (MMC) lead to additional computational advantages due to the application of less complicated constitutive laws of the phases being J_2-plasticity or even linear elasticity (e.g., Pettermann et al. (2010)).

Very recently, the hierarchical multi-scale modeling has been investigated, the idea of which is to have even less computational costs by identifying an anisotropic constitutive law by virtual experiments (Roters et al., 2010; Gawad et al., 2013; Van Houtte et al., 2012). The evolution of the plastic potential according to texture evolution is taken into account by occasionally recomputing the virtual experiments on the microscale if required due to the local state followed by the reidentification of the plastic potential. In the above mentioned works, the virtual experiments are carried out based on the Taylor and ALAMEL approaches.

1.3 Notation and list of frequently used symbols

In this thesis, a direct tensor notation is preferred. In a few cases, the index notation is applied presuming that the summation convention holds. Frequently used symbols and operators are given below.

Abbreviations

CODF	Crystallite orientation distribution function
EBSD	Electron backscatter diffraction
FEM	Finite element method
HS	Hashin-Shtrikman
NL-HS	Non-linear Hashin-Shtrikman type procedure
RD, TD, ND	Rolling, transverse and normal directions of a sheet metal
RVE	Representative volume element
SA	Singular approximation
SC	Self-consistent method

Greek letters

ϵ	Ricci tensor (third-order tensor)
$\dot{\gamma}$	Slip rate
ε	Infinitesimal strain tensor
ω	Infinitesimal spin tensor
σ	Cauchy stress tensor
$\lambda_{1,2,\dots}$	Eigenvalues
τ	Resolved shear stress
$\varphi_1, \Phi, \varphi_2$	Euler angles in 'zxz'-convention

Latin letters

$a, b, t, T, \dots$	Scalar quantities
$\boldsymbol{a}, \boldsymbol{b}, \boldsymbol{c}, \dots$	First-order tensors
$\boldsymbol{A}, \boldsymbol{B}, \boldsymbol{C}, \dots$	Second-order tensors
$\mathbb{A}, \mathbb{B}, \mathbb{C}, \dots$	Fourth-order tensors
$\boldsymbol{F}$	Deformation gradient
$\mathbb{A}, \mathbb{B}$	Fourth-order strain and stress localization tensors
$\mathbb{C}, \mathbb{S}$	Fourth-order stiffness and compliance tensors
$\mathbb{C}_0$	Stiffness of the comparison medium
$\mathbb{I}$	Fourth-order identity
$\mathbb{L}$	Hill's constraint tensor
$\mathbb{P}_0$	Hill's polarization tensor
$\mathbb{P}_{1,2,\dots}$	Fourth-order projectors
$\mathbb{V}'$	Fourth-order texture coefficient
$\boldsymbol{H}$	Displacement gradient
$\boldsymbol{I}$	Second-order identity
$\boldsymbol{Q}$	Orthogonal tensor
c_α	Volume fraction of phase α

Sub- and superscripts

$(\cdot)'$	Deviatoric quantity
$(\cdot)^A$	Skew-symmetric quantity

$(\cdot)^C$	Cubic quantity
$(\cdot)^I$	Isotropic quantity
$(\cdot)^S$	Symmetric quantity
$(\cdot)_{\langle\alpha\rangle}$	Tensor of rank α
$\bar{(\cdot)}$	Effective (macroscopic) quantity
$\hat{(\cdot)}$	Quantity in Fourier space
$\tilde{(\cdot)}$	Fluctuating part of a quantity
$(\cdot)^{\mathsf{T_L}}$	Transposition of left index pair (left minor transposition)
$(\cdot)^{\mathsf{T_R}}$	Transposition of right index pair (right minor transposition)
$(\cdot)^{\mathsf{T}}$	Transposition, major transposition for higher-order tensors of even rank

Tensor operations

$[[\cdot]]$	Special contraction $(a \otimes b) \cdot (\mathbb{C}[[a \otimes b]]) = (a \otimes a) \cdot (\mathbb{C}[b \otimes b])$
$[\cdot]$	Linear mapping $A = \mathbb{C}[B] \;\hat{=}\; C_{ijkl}B_{kl}$
$\square$	Box product, special dyadic product $(A\square B)[C] = ACB$
$\cdot$	Dot product $A \cdot B = A_{ij}B_{ij}$
$\det(\cdot)$	Determinant
$\mathrm{div}\,(\cdot)$	Divergence of a tensor
$\mathrm{Grad}\,(\cdot)$	Lagrangian gradient of a quantity
$\mathrm{grad}\,(\cdot)$	Eulerean gradient of a quantity
$\langle\cdot\rangle$	Volume/ensemble average of a quantity or Maccauley bracket for determining slip activity
$\mathrm{skw}(\cdot)$	Skew-symmetric part
$\mathrm{sym}(\cdot)$	Symmetric part
$\otimes$	Dyadic product $A \otimes B \;\hat{=}\; A_{ij}B_{kl}$
$\star$	Rayleigh product: $Q \star A = A_{ij}(Q e_i) \otimes (Q e_j)$
$\mathrm{tr}(\cdot)$	Trace of a tensor

Chapter 2

Theoretical Foundations

2.1 Kinematics and balance equations

This section provides the main equations being essential for this work. A complete description of the continuum theory can be found in basic text books as, e.g., Betten (1993); Bertram (2005).

2.1.1 Kinematics

The position of each material point of a three-dimensional body at any time is described by the position vector

$$\boldsymbol{x} = \chi(\boldsymbol{X}, t),\tag{2.1}$$

where $\boldsymbol{X}$ is the position of the same point in the reference placement of the body $\mathcal{B}_0$ at time $t = t_0$ (Fig. 2.1). The inverse mapping to identify the position of a point known in the current placement of the body $\mathcal{B}_t$ in the reference frame is defined by $\boldsymbol{X} = \chi^{-1}(\boldsymbol{x}, t)$.

The difference of the position vectors of a material point in the reference and the current placement is called displacement, or more general, in case of all points of the body displacement field

$$\boldsymbol{u}(\boldsymbol{X}, t) = \boldsymbol{x} - \boldsymbol{X} = \chi(\boldsymbol{X}, t) - \boldsymbol{X}.\tag{2.2}$$

For time derivatives of a function $\phi = \phi_L(\boldsymbol{X}, t) = \phi_E(\boldsymbol{x}, t)$, attention has to be paid to the placement, the field is defined on. In the first case, $\phi = \phi_L(\boldsymbol{X}, t)$ is defined in terms of the reference placement being called the Lagrangean description. On the other hand, $\phi = \phi_E(\boldsymbol{x}, t)$ is given with respect to the

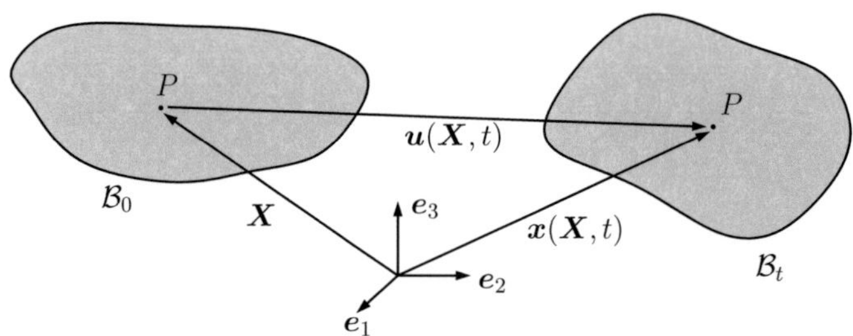

Figure 2.1: Reference $\mathcal{B}_0$ and current placement $\mathcal{B}_t$ of a body with different locations of the same material point P

current placement being the Eulerean description. Therefore, the material time derivative denoted by $\dot{\phi}$ is defined as

$$\dot{\phi} = \frac{\partial \phi_L(\boldsymbol{X}, t)}{\partial t} = \frac{\partial \phi_E(\boldsymbol{x}, t)}{\partial t} + \frac{\partial \phi_E(\boldsymbol{x}, t)}{\partial \boldsymbol{x}} \cdot \dot{\boldsymbol{x}}, \tag{2.3}$$

where $\dot{\boldsymbol{x}} = \boldsymbol{v}(\boldsymbol{x}, t)$ is the velocity of the material point. Note that the mathematical operation between the two partial derivatives denoted by a dot in Eq. (2.3) is a single contraction, so that the above definition holds for any tensorial rank of field ϕ in general.

Spatial derivatives of $\boldsymbol{x}$ and $\boldsymbol{u}$ are used to describe the deformation of the body. The deformation gradient $\boldsymbol{F}$ and the displacement gradient $\boldsymbol{H}$ are defined by

$$\boldsymbol{F} = \mathrm{Grad}\,(\boldsymbol{x}) = \frac{\partial \boldsymbol{x}(\boldsymbol{X}, t)}{\partial \boldsymbol{X}}, \quad \boldsymbol{H} = \mathrm{Grad}\,(\boldsymbol{u}) = \frac{\partial \boldsymbol{u}(\boldsymbol{X}, t)}{\partial \boldsymbol{X}} = \boldsymbol{F} - \boldsymbol{I}, \tag{2.4}$$

therefore, being Lagrangean gradients. $\boldsymbol{I}$ is the second-order identity. Although $\boldsymbol{F}$ is called deformation gradient, it should be kept in mind that it also accounts for rigid body rotations.

The velocity gradient is a Eulerean gradient defined by

$$\boldsymbol{L} = \mathrm{grad}\,(\boldsymbol{v}) = \frac{\partial \boldsymbol{v}(\boldsymbol{x}, t)}{\partial \boldsymbol{x}} = \dot{\boldsymbol{F}} \boldsymbol{F}^{-1}, \tag{2.5}$$

and can be additively decomposed into a symmetric and asymmetric part, $\boldsymbol{D} = \mathrm{sym}(\boldsymbol{L})$ and $\boldsymbol{W} = \mathrm{skw}(\boldsymbol{L})$, respectively, with $\boldsymbol{D}$ being the tensor of rate of deformation and $\boldsymbol{W}$ the spin tensor.

A deformation measure in the reference placement is Green's strain tensor, describing the change of length and angles of line elements in the body

$$\boldsymbol{E} = \frac{1}{2}\left(\boldsymbol{F}^{\mathsf{T}}\boldsymbol{F} - \boldsymbol{I}\right) = \frac{1}{2}\left(\boldsymbol{H} + \boldsymbol{H}^{\mathsf{T}} + \boldsymbol{H}^{\mathsf{T}}\boldsymbol{H}\right), \tag{2.6}$$

the linearization of which is the infinitesimal strain tensor

$$\varepsilon = \frac{1}{2}\left(\boldsymbol{H} + \boldsymbol{H}^{\mathsf{T}}\right) = \mathrm{sym}(\boldsymbol{H}). \tag{2.7}$$

The skew-symmetric part of $\boldsymbol{H}$ is the infinitesimal rotation $\omega = \left(\boldsymbol{H} - \boldsymbol{H}^{\mathsf{T}}\right)/2 = \mathrm{skw}(\boldsymbol{H})$.

It should be noted that there exist several strain measures as, e.g., Biot strain, Almansi strain and Hencky strain, the definitions of which are omitted for brevity in this thesis. More details are to be found in basic continuum mechanics literature as, e.g., Truesdell and Noll (1965); Bertram (2005).

Every second-order tensor can be uniquely split up into a spherical and a deviatoric part. The deviatoric part can even be furthermore decomposed into a symmetric and skew-symmetric part. In case of infinitesimal strain, the spherical part ε° describes the volumetric change of the body (dilatation) and the deviatoric one ε' (this is always symmetric in case of strain) the volume preserving change of the shape of the body (distortion)

$$\varepsilon = \varepsilon^\circ + \varepsilon'. \tag{2.8}$$

Two isotropic projectors $\mathbb{P}_1^I = (\boldsymbol{I} \otimes \boldsymbol{I})/3$ and $\mathbb{P}_2^I = \mathbb{I}^S - \mathbb{P}_1^I$ (see, e.g, Rychlewski and Zhang (1989)) can be introduced, which map each second-order tensor into its spherical and (symmetric) deviatoric part, respectively. $\mathbb{I}^S$ denotes the fourth-order identity on symmetric second-order tensors. In case of the infinitesimal strain, this reads

$$\varepsilon^\circ = \mathbb{P}_1^I[\varepsilon], \quad \varepsilon' = \mathbb{P}_2^I[\varepsilon]. \tag{2.9}$$

2.1.2 Balance of momentum and moment of momentum

In continuum theory, the balance of momentum defines the time derivative of momentum of the body being equal to the external loads, which are usually classified into surface and volumetric loads. Constituting that the balance of mass holds, the balance of momentum for the body reads

$$\frac{\partial}{\partial t}\int_B \varrho\boldsymbol{v}\,\mathrm{d}V = \int_{\partial B} \boldsymbol{t}\,\mathrm{d}A + \int_B \varrho\boldsymbol{b}\,\mathrm{d}V, \tag{2.10}$$

with the traction $\boldsymbol{t}$ on the boundary ∂B with normal $\boldsymbol{n}$ and body force $\boldsymbol{b}$ as well as the mass density ϱ. Using Gauss' theorem to transform the surface integral into a volume integral and stating that the balance of momentum should hold for each body and all sub-bodies, the local form of this balance equation reads

$$\mathrm{div}\,(\boldsymbol{\sigma}) + \varrho \boldsymbol{b} = \varrho \dot{\boldsymbol{v}}. \tag{2.11}$$

The divergence of a field $\boldsymbol{f}(\boldsymbol{x})$ is defined as

$$\mathrm{div}\,(\boldsymbol{f}(\boldsymbol{x})) = \mathrm{grad}\,(\boldsymbol{f}(\boldsymbol{x})) \cdot \cdot \boldsymbol{I}. \tag{2.12}$$

Note that in Eq. (2.12) the two dots denote the double contraction with the second-order identity so that this definition holds for all tensorial ranks of field $\boldsymbol{f}$.

Since throughout this work, only static equilibrium is to be considered, Eq. (2.11) reads for the static case

$$\mathrm{div}\,(\boldsymbol{\sigma}) + \varrho \boldsymbol{b} = \boldsymbol{o}, \tag{2.13}$$

with the zero vector $\boldsymbol{o}$.

Furthermore, boundary conditions on the boundary $\partial \mathcal{B}$ of $\mathcal{B}$ are to be added. Therefore, the boundary of $\mathcal{B}$ is divided into a Neumann $\partial \mathcal{B}^{\sigma}$ and Dirichlet $\partial \mathcal{B}^{u}$ boundary, on which the traction $\boldsymbol{t}$ and displacement $\boldsymbol{u}$ is prescribed, respectively,

$$\boldsymbol{t} = \boldsymbol{\sigma} \boldsymbol{n} = \tilde{\boldsymbol{t}} \quad \mathrm{on} \quad \partial \mathcal{B}^{\sigma}, \tag{2.14}$$

$$\boldsymbol{u} = \tilde{\boldsymbol{u}} \quad \mathrm{on} \quad \partial \mathcal{B}^{u}, \tag{2.15}$$

$$\partial \mathcal{B}^{\sigma} \cap \partial \mathcal{B}^{u} = \emptyset, \quad \partial \mathcal{B}^{\sigma} \cup \partial \mathcal{B}^{u} = \partial \mathcal{B}. \tag{2.16}$$

Additionally assuming that pure distributions of moments are absent in and on the body, the balance equation of moment of momentum requests the symmetry of the stress tensor

$$\boldsymbol{\sigma} = \boldsymbol{\sigma}^{\mathsf{T}}. \tag{2.17}$$

2.1.3 Dissipation inequality

The second law of thermodynamics states that entropy production is never negative (see textbooks as, e.g., Haupt (2000)). The local form of this entropy inequality is given in the form

$$p_{\eta} = \varrho \dot{\eta} - \frac{\varrho w}{\Theta} + \mathrm{div}\left(\frac{\boldsymbol{q}}{\Theta}\right) \geq 0, \tag{2.18}$$

where η is the specific entropy, p_η is the production of entropy, Θ is the temperature, w/Θ is the density of volume distributed entropy supply and $\boldsymbol{q}$ is the heat flux. Furthermore, the local form of the balance of specific internal energy e is given by

$$\varrho\dot{e} = \varrho w + \boldsymbol{\sigma} \cdot \dot{\boldsymbol{\varepsilon}} - \operatorname{div}\left(\boldsymbol{q}\right). \tag{2.19}$$

Using the relation of specific internal energy and specific free energy ψ

$$\psi = e - \eta\Theta \tag{2.20}$$

and its time derivative, taking into account that in general $\psi = \psi(\boldsymbol{\varepsilon}, \Theta, \boldsymbol{g} = \operatorname{grad}(\Theta))$

$$\dot{\psi} = \frac{\partial\psi}{\partial\boldsymbol{\varepsilon}} \cdot \dot{\boldsymbol{\varepsilon}} + \frac{\partial\psi}{\partial\Theta}\dot{\Theta} + \frac{\partial\psi}{\partial\boldsymbol{g}} \cdot \dot{\boldsymbol{g}} = \dot{e} - \dot{\Theta}\eta - \Theta\dot{\eta}, \tag{2.21}$$

the Eqs. (2.18), (2.19) and (2.21) can be combined to get

$$\Theta p_\eta = \left(\boldsymbol{\sigma} - \varrho\frac{\partial\psi}{\partial\boldsymbol{\varepsilon}}\right) \cdot \dot{\boldsymbol{\varepsilon}} - \varrho\left(\eta + \frac{\partial\psi}{\partial\Theta}\right)\dot{\Theta} - \varrho\frac{\partial\psi}{\partial\boldsymbol{g}} \cdot \dot{\boldsymbol{g}} - \frac{1}{\Theta}\boldsymbol{q} \cdot \boldsymbol{g} \geq 0. \tag{2.22}$$

The latter equation is called Clausius-Duhem-inequality or also dissipation inequality, since Θp_η is the internal dissipation. Eq. (2.22) needs to hold for all thermo-kinematical processes with arbitrary $\dot{\boldsymbol{\varepsilon}}$, $\dot{\Theta}$ and $\dot{\boldsymbol{g}}$, so that the following potential relations can be derived

$$\boldsymbol{\sigma} = \varrho\frac{\partial\psi}{\partial\boldsymbol{\varepsilon}}, \quad \eta = -\frac{\partial\psi}{\partial\Theta}. \tag{2.23}$$

Additionally, it can be concluded that $\psi = \psi(\boldsymbol{\varepsilon}, \Theta) \neq \psi(\boldsymbol{g})$. The remaining term $-\boldsymbol{q} \cdot \boldsymbol{g}/\Theta \geq 0$ can be interpreted, when using Fouriers law of heat conduction $\boldsymbol{q} = -\boldsymbol{\kappa}\boldsymbol{g}$ with $\boldsymbol{\kappa}$ being the generally anisotropic second-order tensor of coefficients of heat conductivity, in a sense that $\boldsymbol{\kappa}$ is positive semi-definite, since $\boldsymbol{g} \cdot \boldsymbol{\kappa}\boldsymbol{g}/\Theta \geq 0$.

Note that the dot in Eqs. (2.21) and (2.22) and in the sequel is interpreted as dot product between two tensors of the same but arbitrary order.

2.2 Constitutive law of single crystals with cubic crystal symmetry

2.2.1 Geometrically linear framework for elasticity

Using the geometrically linear or small strain framework, it is not necessary to distinguish between the reference and the current placement of a body.

Therefore, all variables are defined with respect to the single existing placement of the body $\mathcal{B}$. In this case, the Cauchy stress $\boldsymbol{\sigma}$ and the infinitesimal strain tensor $\boldsymbol{\varepsilon}$ define the elastic law

$$\boldsymbol{\sigma} = \mathbb{C}[\boldsymbol{\varepsilon}] \qquad \boldsymbol{\varepsilon} = \mathbb{S}[\boldsymbol{\sigma}]. \tag{2.24}$$

The square brackets $[\cdot]$ denote the linear mapping, i.e. $\mathbb{A}[\boldsymbol{B}] \mathrel{\hat{=}} A_{ijkl} B_{kl}$. The stiffness tensor $\mathbb{C}$ as well as the compliance tensor $\mathbb{S} = \mathbb{C}^{-1}$ possess a certain symmetry $\mathcal{S}$

$$\mathbb{C} = \boldsymbol{R} \star \mathbb{C} \qquad \mathbb{S} = \boldsymbol{R} \star \mathbb{S} \qquad \forall \boldsymbol{R} \in \mathcal{S}, \tag{2.25}$$

which primarily is the group of cubic symmetry in the present work.

Therefore, the aforementioned elasticity tensors can be denoted in the projector representation (Halmos, 1958; Bertram and Olschewski, 1991; Rychlewski and Zhang, 1989; Bertram, 2005)

$$\mathbb{C} = \lambda_1 \mathbb{P}_1^C + \lambda_2 \mathbb{P}_2^C + \lambda_3 \mathbb{P}_3^C \qquad \mathbb{S} = \frac{1}{\lambda_1} \mathbb{P}_1^C + \frac{1}{\lambda_2} \mathbb{P}_2^C + \frac{1}{\lambda_3} \mathbb{P}_3^C, \tag{2.26}$$

with the three cubic eigenvalues being composed of the three elastic constants $\lambda_1 = C_{1111} + 2C_{1122}$, $\lambda_2 = C_{1111} - C_{1122}$ and $\lambda_3 = 2C_{1212}$ as well as the three cubic projectors

$$\mathbb{P}_1^C = \frac{1}{3} \boldsymbol{I} \otimes \boldsymbol{I}, \qquad \mathbb{P}_2^C = \mathbb{D} - \mathbb{P}_1^C, \qquad \mathbb{P}_3^C = \mathbb{I}^S - \mathbb{D}, \tag{2.27}$$

with $\mathbb{I}^S$ being the fourth-order identity on symmetric second-order tensors. The second and third cubic projectors incorporate the crystal orientation $\boldsymbol{Q} = \boldsymbol{g}_i \otimes \boldsymbol{e}_i \in SO(3)$ of the single crystal with the basis vectors of the reference and crystal coordinate system $\boldsymbol{e}_i$ and $\boldsymbol{g}_i$, $i = 1, 2, 3$, respectively. The crystal orientation $\boldsymbol{Q}$ maps the fixed reference system onto the crystal lattice vectors and, therefore, is used to rotate an anisotropic fourth-order tensor $\mathbb{D}_0$ given in terms of sample coordinates by the Rayleigh product $\star$ to the particular crystal coordinate system

$$\mathbb{D} = \boldsymbol{Q} \star \mathbb{D}_0 = \boldsymbol{Q} \star \sum_{i=1}^{3} \boldsymbol{e}_i \otimes \boldsymbol{e}_i \otimes \boldsymbol{e}_i \otimes \boldsymbol{e}_i = \sum_{i=1}^{3} \boldsymbol{g}_i \otimes \boldsymbol{g}_i \otimes \boldsymbol{g}_i \otimes \boldsymbol{g}_i. \tag{2.28}$$

2.2.2 Geometrically linear framework for elasto-viscoplasticity

In the general case of small strain inelasticity, the elastic law reads

$$\boldsymbol{\sigma} = \mathbb{C}[\boldsymbol{\varepsilon}_e], \quad \boldsymbol{\varepsilon} = \boldsymbol{\varepsilon}_e + \boldsymbol{\varepsilon}_p, \tag{2.29}$$

with the additive decomposition of the infinitesimal strain tensor into an elastic and an inelastic part, ε_e and ε_p, respectively. The flow rule is given by an evolution equation for the inelastic part of the strain tensor

$$\dot{\varepsilon}_p = \sum_{\alpha=1}^{12} \dot{\gamma}_\alpha \mathrm{sym}(\boldsymbol{d}_\alpha \otimes \boldsymbol{n}_\alpha), \tag{2.30}$$

with the directions $\boldsymbol{d}_\alpha$ and the normals $\boldsymbol{n}_\alpha$ of the octahedral slip systems of the face-centered cubic crystal. The evolution of slip in the αth slip system is described by the overstress relation (e.g. Méric et al. (1994))

$$\dot{\gamma}_\alpha = \dot{\gamma}_0 \, \mathrm{sgn}\,(\tau_\alpha) \left\langle \frac{|\tau_\alpha| - \tau_C}{\tau_D} \right\rangle^m, \tag{2.31}$$

with the Schmid stresses $\tau_\alpha = \boldsymbol{\sigma} \cdot \boldsymbol{M}_\alpha$ and $\boldsymbol{M}_\alpha = \boldsymbol{d}_\alpha \otimes \boldsymbol{n}_\alpha$ as well as the critical resolved shear stress τ_C that is assumed equal for all slip systems. Furthermore, τ_D is the drag stress and $\dot{\gamma}_0$ is a reference slip rate. The Macauley brackets $\langle \cdot \rangle$ are defined as $\langle x \rangle = \max(0, x)$. A Voce-type exponential law is applied to describe the hardening behavior

$$\tau_C = \tau_{C0} + (\tau_{C\infty} - \tau_{C0}) \left(1 - \exp\left(-\frac{\Theta_0 \gamma}{\tau_{C\infty} - \tau_{C0}} \right) \right), \tag{2.32}$$

with the initial and asymptotic values for the critical resolved shear stress, τ_{C0} and $\tau_{C\infty}$, respectively. Θ_0 is the initial hardening modulus and $\dot{\gamma} = \sum_{\alpha=1}^{12} |\dot{\gamma}_\alpha|$ is the accumulated slip. In the present approach, kinematic hardening is neglected assuming vanishing back stresses so that the hardening law in Eq. (2.32) describes purely isotropic hardening behavior.

After obtaining the slip rates, the lattice rotation $\boldsymbol{Q}$ can be explicitly updated

$$\dot{\boldsymbol{\omega}}_e = \dot{\boldsymbol{\omega}} - \dot{\boldsymbol{\omega}}_p, \quad \dot{\boldsymbol{\omega}}_p = \sum_{\alpha=1}^{12} \dot{\gamma}_\alpha \mathrm{skw}(\boldsymbol{d}_\alpha \otimes \boldsymbol{n}_\alpha), \quad \dot{\boldsymbol{Q}}\boldsymbol{Q}^{-1} = \boldsymbol{\omega}_e, \tag{2.33}$$

so as to modify the rotation of the stiffness tensor $\mathbb{C}$ and the Schmid tensors $\boldsymbol{M}_\alpha$ in each time step according to the deformation.

2.3 Representation of crystal orientations and orientation distributions

For the description of polycrystalline material behavior by the single crystalline constitutive law, the orientation of each crystal plays a crucial role. In materials science it is widely spread to use three Euler angles (Bunge, 1993), or Miller indices to parametrize the orientation tensor. Other common possibilities to describe the position of each individual crystal with respect to the sample frame are the usage of rotation axis/angle or quaternions. A detailed overview of parametrizations and the relationship between different methods are given in Hansen et al. (1978); Bunge (1993); Adams and Olson (1998); Morawiec (2004).

2.3.1 Euler angle parametrization

It is already known for nearly three centuries that three parameters are sufficient to properly describe the orientation of, e.g., a crystal, in space. A common parametrization is the usage of Euler angles, the definition of which depends on the application. In aeronautics, a definition called the $'zyx$-convention' is applied, whereas materials scientists usually use the $'zxz$-convention' which is also used throughout this work. The rotation of an object described in the reference frame is, therefore, first rotated around the reference z-axis, followed by a rotation about the new x-axis and finally rotated about the new z-axis. The angles describing these individual rotations are usually named $\varphi_1, \Phi, \varphi_2$, respectively. If the rotation tensor is expressed by

$$\boldsymbol{Q} = \boldsymbol{g}_i \otimes \boldsymbol{e}_i, \tag{2.34}$$

the axis of the resulting rotated coordinate system with respect to the reference system are given in columns in the matrix representation of $\boldsymbol{Q}$. The components of the individual rotations $\boldsymbol{Q}_z(\alpha)$ and $\boldsymbol{Q}_x(\beta)$ by angles α and β with respect to the reference coordinate system (sample system) are given as

$$Q_{ij}^z(\alpha) = \begin{pmatrix} \cos(\alpha) & -\sin(\alpha) & 0 \\ \sin\alpha & \cos(\alpha) & 0 \\ 0 & 0 & 1 \end{pmatrix}, \quad Q_{ij}^x(\beta) = \begin{pmatrix} 1 & 0 & 0 \\ 0 & \cos(\beta) & -\sin(\beta) \\ 0 & \sin(\beta) & \cos(\beta) \end{pmatrix}, \tag{2.35}$$

so that the resulting zxz-rotation is given by

$$\boldsymbol{Q} = \boldsymbol{Q}_z(\varphi_1)\boldsymbol{Q}_x(\Phi)\boldsymbol{Q}_z(\varphi_2), \tag{2.36}$$

with components with respect to the reference system

$$Q_{ij}(\varphi_1, \Phi, \varphi_2) =$$

$$\begin{pmatrix} \cos(\varphi_1)\cos(\varphi_2) - \sin(\varphi_1)\cos(\Phi)\sin(\varphi_2) & -\cos(\varphi_1)\sin(\varphi_2) - \sin(\varphi_1)\cos(\Phi)\cos(\varphi_2) & \sin(\varphi_1)\sin(\Phi) \\ \sin(\varphi_1)\cos(\varphi_2) + \cos(\varphi_1)\cos(\Phi)\sin(\varphi_2) & -\sin(\varphi_1)\sin(\varphi_2) + \cos(\varphi_1)\cos(\Phi)\cos(\varphi_2) & -\cos(\varphi_1)\sin(\Phi) \\ \sin(\Phi)\sin(\varphi_2) & \sin(\Phi)\cos(\varphi_2) & \cos(\Phi) \end{pmatrix}.$$

$$(2.37)$$

2.3.2 Crystallographically equivalent orientations

Due to the definition of the orientation tensor in Eq. (2.34) with the components given by Eq. (2.37), which is the transpose of the definition given by Bunge (1993), the symmetrically equivalent orientations reflecting crystal and sample symmetry are determined as

$$\boldsymbol{Q}^{eq} = \boldsymbol{Q}^S \boldsymbol{Q} \boldsymbol{Q}^C, \quad \forall \boldsymbol{Q}^C \in S^C \subseteq SO(3), \quad \forall \boldsymbol{Q}^S \in S^S \subseteq SO(3). \qquad (2.38)$$

S^C and S^S denote the symmetry group of the crystal and sample, respectively. In this work, polycrystals consisting of single crystals with cubic crystal symmetry are to be examined. For cubic symmetry, 24 symmetrically equivalent orientations exist. When thinking of a rotation in terms of a rotational axis and an angle, these 24 symmetry transformations are generated by rotating with angle $\pi/2$ and π around the cube edges, with angle π around the face diagonals of the cube and with $2\pi/3$ around the space diagonals. The components of the orthogonal tensors of these cubic symmetry transformations are given in Tab. 2.1. Furthermore, if the polycrystal is somehow pretreated, as, e.g., by rolling a sheet metal, additionally a particular sample symmetry shows to be observable. In case of rolling, this is the orthotropic sample symmetry with additional four equivalent symmetry operations being rotations with angle π around the particular axes of the rolled sheet metal, being the rolling, transverse and normal directions, respectively. The orthotropic symmetry transformations are given in Tab. 2.2. Altogether, the crystal orientations of a specimen possessing cubic crystal and orthotropic sample symmetry can be equivalently described by 96 orthogonal tensors.

$\boldsymbol{Q}_1^{cub}$	$\boldsymbol{Q}_2^{cub}$	$\boldsymbol{Q}_3^{cub}$	$\boldsymbol{Q}_4^{cub}$
$\begin{pmatrix} 1 & 0 & 0 \\ 0 & 1 & 0 \\ 0 & 0 & 1 \end{pmatrix}$	$\begin{pmatrix} -1 & 0 & 0 \\ 0 & 1 & 0 \\ 0 & 0 & -1 \end{pmatrix}$	$\begin{pmatrix} -1 & 0 & 0 \\ 0 & -1 & 0 \\ 0 & 0 & 1 \end{pmatrix}$	$\begin{pmatrix} 1 & 0 & 0 \\ 0 & 0 & -1 \\ 0 & 1 & 0 \end{pmatrix}$
$\boldsymbol{Q}_5^{cub}$	$\boldsymbol{Q}_6^{cub}$	$\boldsymbol{Q}_7^{cub}$	$\boldsymbol{Q}_8^{cub}$
$\begin{pmatrix} 0 & 0 & 1 \\ 0 & 1 & 0 \\ -1 & 0 & 0 \end{pmatrix}$	$\begin{pmatrix} 0 & -1 & 0 \\ 1 & 0 & 0 \\ 0 & 0 & 1 \end{pmatrix}$	$\begin{pmatrix} 1 & 0 & 0 \\ 0 & 0 & 1 \\ 0 & -1 & 0 \end{pmatrix}$	$\begin{pmatrix} 0 & 0 & -1 \\ 0 & 1 & 0 \\ 1 & 0 & 0 \end{pmatrix}$
$\boldsymbol{Q}_9^{cub}$	$\boldsymbol{Q}_{10}^{cub}$	$\boldsymbol{Q}_{11}^{cub}$	$\boldsymbol{Q}_{12}^{cub}$
$\begin{pmatrix} 0 & 1 & 0 \\ -1 & 0 & 0 \\ 0 & 0 & 1 \end{pmatrix}$	$\begin{pmatrix} 0 & 1 & 0 \\ 1 & 0 & 0 \\ 0 & 0 & -1 \end{pmatrix}$	$\begin{pmatrix} 0 & -1 & 0 \\ -1 & 0 & 0 \\ 0 & 0 & -1 \end{pmatrix}$	$\begin{pmatrix} 0 & 0 & 1 \\ 0 & -1 & 0 \\ 1 & 0 & 0 \end{pmatrix}$
$\boldsymbol{Q}_{13}^{cub}$	$\boldsymbol{Q}_{14}^{cub}$	$\boldsymbol{Q}_{15}^{cub}$	$\boldsymbol{Q}_{16}^{cub}$
$\begin{pmatrix} 0 & 0 & -1 \\ 0 & -1 & 0 \\ -1 & 0 & 0 \end{pmatrix}$	$\begin{pmatrix} -1 & 0 & 0 \\ 0 & 0 & 1 \\ 0 & 1 & 0 \end{pmatrix}$	$\begin{pmatrix} -1 & 0 & 0 \\ 0 & 0 & -1 \\ 0 & -1 & 0 \end{pmatrix}$	$\begin{pmatrix} 0 & 0 & 1 \\ 1 & 0 & 0 \\ 0 & 1 & 0 \end{pmatrix}$
$\boldsymbol{Q}_{17}^{cub}$	$\boldsymbol{Q}_{18}^{cub}$	$\boldsymbol{Q}_{19}^{cub}$	$\boldsymbol{Q}_{20}^{cub}$
$\begin{pmatrix} 0 & -1 & 0 \\ 0 & 0 & 1 \\ -1 & 0 & 0 \end{pmatrix}$	$\begin{pmatrix} 0 & -1 & 0 \\ 0 & 0 & -1 \\ 1 & 0 & 0 \end{pmatrix}$	$\begin{pmatrix} 0 & 1 & 0 \\ 0 & 0 & -1 \\ -1 & 0 & 0 \end{pmatrix}$	$\begin{pmatrix} 0 & 0 & -1 \\ 1 & 0 & 0 \\ 0 & -1 & 0 \end{pmatrix}$
$\boldsymbol{Q}_{21}^{cub}$	$\boldsymbol{Q}_{22}^{cub}$	$\boldsymbol{Q}_{23}^{cub}$	$\boldsymbol{Q}_{24}^{cub}$
$\begin{pmatrix} 0 & 0 & 1 \\ -1 & 0 & 0 \\ 0 & -1 & 0 \end{pmatrix}$	$\begin{pmatrix} 0 & 0 & -1 \\ -1 & 0 & 0 \\ 0 & 1 & 0 \end{pmatrix}$	$\begin{pmatrix} 0 & 1 & 0 \\ 0 & 0 & 1 \\ 1 & 0 & 0 \end{pmatrix}$	$\begin{pmatrix} 1 & 0 & 0 \\ 0 & -1 & 0 \\ 0 & 0 & -1 \end{pmatrix}$

Table 2.1: Equivalent symmetry transformations of cubic symmetry group

$\boldsymbol{Q}_1^{orth}$	$\boldsymbol{Q}_2^{orth}$	$\boldsymbol{Q}_3^{orth}$	$\boldsymbol{Q}_4^{orth}$
$\begin{pmatrix} 1 & 0 & 0 \\ 0 & 1 & 0 \\ 0 & 0 & 1 \end{pmatrix}$	$\begin{pmatrix} -1 & 0 & 0 \\ 0 & 1 & 0 \\ 0 & 0 & -1 \end{pmatrix}$	$\begin{pmatrix} -1 & 0 & 0 \\ 0 & -1 & 0 \\ 0 & 0 & 1 \end{pmatrix}$	$\begin{pmatrix} 1 & 0 & 0 \\ 0 & -1 & 0 \\ 0 & 0 & -1 \end{pmatrix}$

Table 2.2: Equivalent symmetry transformations of orthotropic symmetry group

2.3.3 Crystallite orientation distribution function

In order to describe the orientation distribution in a polycrystalline aggregate, the crystallite orientation distribution function (CODF) is introduced. The CODF $f(\boldsymbol{Q})$ specifies the volume fraction $\mathrm{d}v/v$ of crystals with the orientation $\boldsymbol{Q} + \mathrm{d}\boldsymbol{Q}$ (Bunge, 1965, 1968a), i.e.

$$\frac{\mathrm{d}v}{v}(\boldsymbol{Q}) = f(\boldsymbol{Q})\,\mathrm{d}Q. \tag{2.39}$$

$\mathrm{d}Q$ is the volume element in $SO(3)$. The function $f(\boldsymbol{Q})$ is non-negative and normalized

$$\int_{SO(3)} f(\boldsymbol{Q})\,\mathrm{d}Q = 1. \tag{2.40}$$

The crystallite orientation distribution function $f(\boldsymbol{Q})$ reflects both, the symmetry of the crystallites forming the aggregate and the sample symmetry, which results from the processing history (Bunge, 1993) (cf. section 2.3.2). The crystal symmetry implies the following symmetry relation: $f(\boldsymbol{Q}) = f(\boldsymbol{Q}\boldsymbol{Q}^C)$. The sample symmetry implies $f(\boldsymbol{Q}) = f(\boldsymbol{Q}^S\boldsymbol{Q})$.

The CODF can be represented by a Fourier expansion. For the special case of a cubic crystal symmetry, this expansion has been used by several authors (Adams et al., 1992; Guidi et al., 1992; Böhlke, 2005, 2006; Böhlke et al., 2010). For aggregates of cubic crystals, the Fourier expansion has the following form

$$f(\boldsymbol{Q}) = 1 + \sum_{i=1}^{\infty} f_{\alpha_i}(\boldsymbol{Q}), \qquad f_{\alpha_i} = \mathbb{V}'_{\langle\alpha_i\rangle} \cdot \mathbb{F}'_{\langle\alpha_i\rangle}(\boldsymbol{Q}), \quad \mathbb{F}'_{\langle\alpha_i\rangle}(\boldsymbol{Q}) = \boldsymbol{Q} \star \mathbb{T}'_{\langle\alpha_i\rangle}, \tag{2.41}$$

with $\alpha_i \in \{4, 6, 8, 9, 10, \ldots\}$. The $\mathbb{V}'_{\langle\alpha_i\rangle}$ are tensorial Fourier coefficients or texture coefficients, where the bracket $\langle\cdot\rangle$ in subscript indicates the tensor rank. The tensors $\mathbb{T}'_{\langle\alpha_i\rangle}$ are reference tensors being normalized without loss of generality $\|\mathbb{T}'_{\langle\alpha_i\rangle}\| = 2\alpha + 1$. All the tensors $\mathbb{V}'_{\langle\alpha_i\rangle}$ and $\mathbb{T}'_{\langle\alpha_i\rangle}$ are irreducible so that they are completely symmetric and traceless. Exemplarily, for the fourth-order texture coefficient, this means

$$V'_{ijkl} = V'_{jikl} = V'_{klij} = V'_{kjil} = \ldots, \qquad V'_{iikl} = 0. \tag{2.42}$$

Due to the aforementioned symmetry relations of the CODF, the reference tensors $\mathbb{T}'_{\langle\alpha_i\rangle}$ reflect the crystal symmetry, i.e. $\mathbb{T}'_{\langle\alpha_i\rangle} = \boldsymbol{Q}^C \star \mathbb{T}'_{\langle\alpha_i\rangle} \forall \boldsymbol{Q}^C \in S^C$, whereas the texture coefficients $\mathbb{V}'_{\langle\alpha_i\rangle}$ reflect the sample symmetry, i.e. $\mathbb{V}'_{\langle\alpha_i\rangle} = \boldsymbol{Q}^S \star \mathbb{V}'_{\langle\alpha_i\rangle} \forall \boldsymbol{Q}^S \in S^S$. For an isotropic microstructure, all texture coefficients are equal to zero. In the single crystalline case, the Frobenius norm of the texture coefficients takes the maximum value, i.e., $\|\mathbb{V}'_{\langle\alpha_i\rangle}\| = 1$.

2.3.4 Visualization of orientation distributions

In this section, two methods to visualize texture data are presented and discussed, namely the pole figures and the orientation space. These methods are frequently used in this work. For a deeper insight or other techniques to visualize texture, e.g., using the Rodrigues space, the reader could refer to, e.g., Bunge (1993); Kocks et al. (1998); Schwartz et al. (2000).

Pole figures and inverse pole figures

The usage of pole figures to represent texture is widely spread in the materials science community. The pole figures are constructed by the stereographic projection, where the intersection point of the normals of the crystallographic planes with the upper hemisphere of a unit sphere is projected onto the equatorial plane using the south pole as reference point for the projection (Fig. 2.2). The stereographic projection preserves angles between to lines but does not preserve areas or distances in contrast to the equal-area and equal-distance projection methods. Due to this fact, a uniform orientation distribution of crystal orientations would show an equal number of poles in equal areas of the unit spheres surface but, simultaneously, shows an apparent clustering in the center of the pole figure when using the stereographic projection.

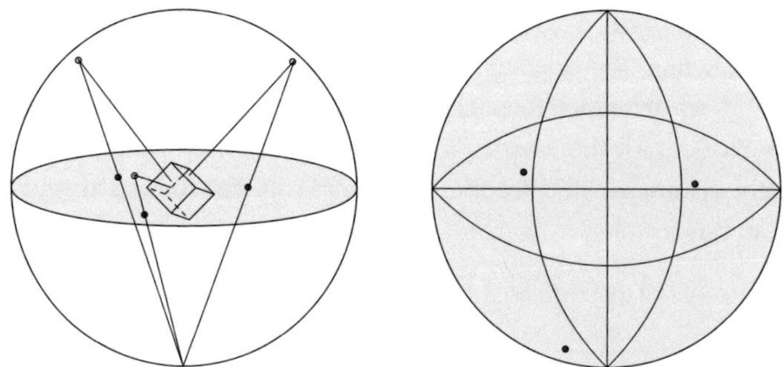

Figure 2.2: Stereographic projection (left) and (100) pole figure (right)

Usually, the projection plane is identical to the sheet metal plane being characterized by the rolling direction (RD) and transverse direction (TD) while the sheet normal is named normal direction (ND). However, the same notation is also used for materials that are not rolled (Schwartz et al., 2009). Fig. 2.3 shows the {100}, {110} and {111} pole figures of a cubic single crystal[1]. Obviously, the {100} pole figure shows three poles, the {110} pole figure six poles and the {111} pole figure four poles for one crystal, the number of poles being identical to the number of crystallographic planes with the respective normals.

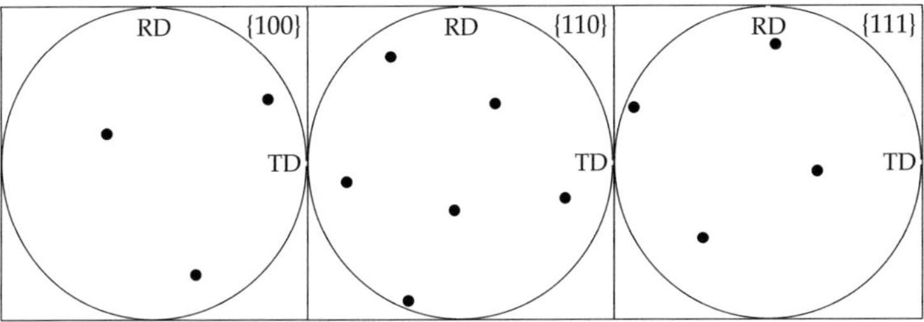

Figure 2.3: Different pole figures of one randomly oriented cubic single crystal

In case of textured materials, the pole figures show clusters of poles meaning an accumulation of crystals having a special orientation. Some typical examples of textures, being the {100} pole figures due to tensile, rolling and shear deformation with about 300 single crystals are shown in Fig. 2.4.

Usually, in materials science, pole figures are not evaluated showing the discrete orientations of each single crystal but are drawn continuously depicting the density distribution of orientations. In this case, each of the poles of the single crystals (Dirac functions) is described by a continuous function, e.g., using smoothing functions of Mises-Fisher or de-la-Vallee-Poussin type (e.g., Schaeben (1997)), the linear combination of which gives a continuous approximation of the orientation distribution in the polycrystal. The continuous orientation distribution function can also be visualized by stereographic projection.

---

[1]Throughout this work, pole figures are generated using the Matlab MTEX Toolbox (http://code.google.com/p/mtex/), where stereographic projection has to be enforced using the 'eangle' option.

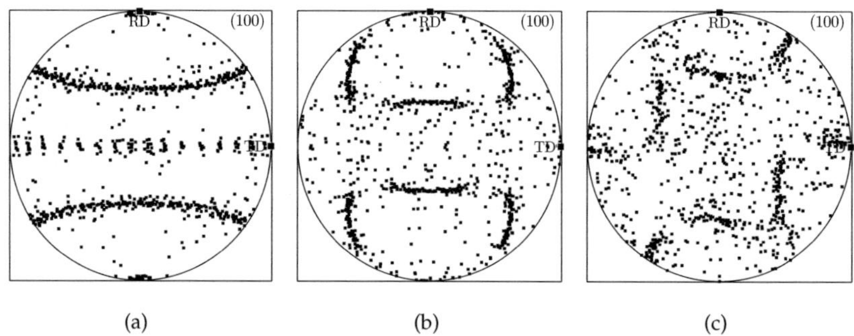

(a)   (b)   (c)

Figure 2.4: Discrete {100} pole figures of (a) tensile, (b) rolling and (c) shear texture

Fig. 2.5 depicts the same textures as Fig. 2.4, showing, however, the continuous orientation distribution. While in pole figures showing discrete poles it is hard to properly illustrate different weights due to the different volume fractions of the crystals, the appearance of the continuous pole figures is strongly influenced by the used kernel function and its parameters to approximate the density. Since, as described above, crystal orientations are projected onto a particular plane of the sample, the symmetry of the sample is observable in the pole figures. For the tensile and rolling textures in Figs. 2.4 and 2.5, an orthorhombic sample symmetry is present, meaning that all quarters of the pole figure contain equivalent information. Nevertheless, usually the complete pole figure is depicted in texture analysis.

In contrast to the above described procedure of generating pole figures, the inverse pole figures are the stereographic projection of particular axis describing the sample symmetry within the local crystallographic coordinate system. Due to this definition, the inverse pole figure reflects the crystal symmetry. Fig. 2.6 shows the complete pole figures of the tensile, rolling and shear texture, respectively. The light lines mark the statistically equivalent regions of the inverse pole figure. Particularly, as shown in Fig. 2.6, for the cubic crystal symmetry, 24 characteristic triangles are separated by these lines. Figs. 2.7 and 2.8 show one of the statistically representative triangles for the above mentioned textures in the discrete and continuous form. In materials science, texture data in inverse pole figures is usually represented by plotting the reduced statistically

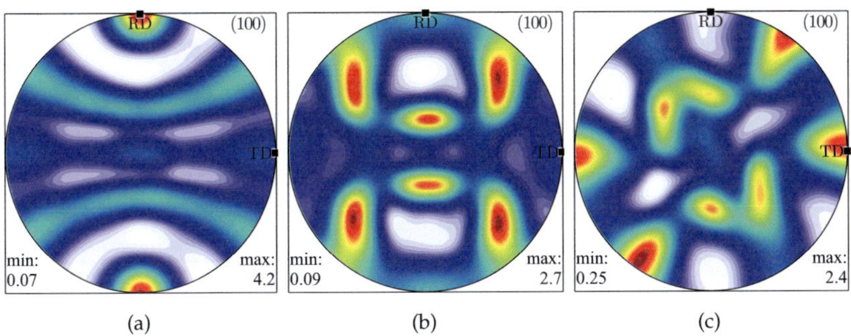

Figure 2.5: Continuous {100} pole figures of (a) tensile, (b) rolling and (c) shear texture

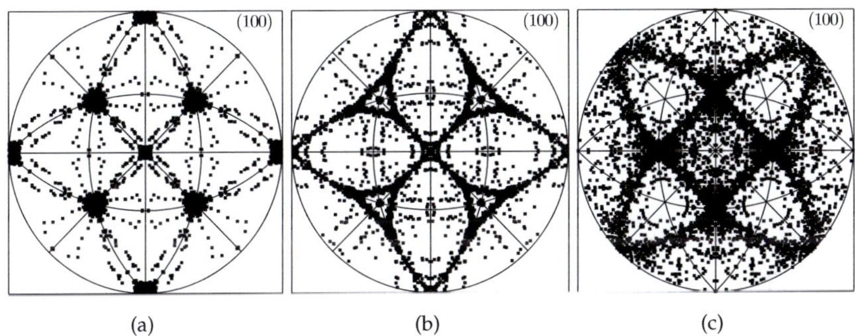

Figure 2.6: Full inverse pole figures of (a) tensile, (b) rolling and (c) shear texture

equivalent part as opposed to the 'normal' pole figures, which are usually shown by their complete circular representation.

One drawback of using pole figures to characterize the texture of the polycrystalline aggregate is the fact that they show only the orientations of characteristic directions or planes of the individual crystals. At least, different pole figures for different crystallographic directions are required to characterize the texture of the specimen, however, the information on which points in different pole figures belong to the same crystal gets lost (e.g., Schwartz et al., 2009).

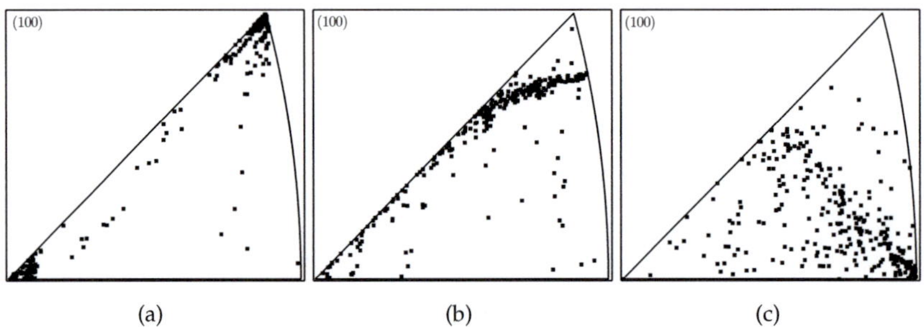

(a)   (b)   (c)

Figure 2.7: Representative triangle of discrete inverse pole figures of (a) tensile, (b) rolling and (c) shear texture

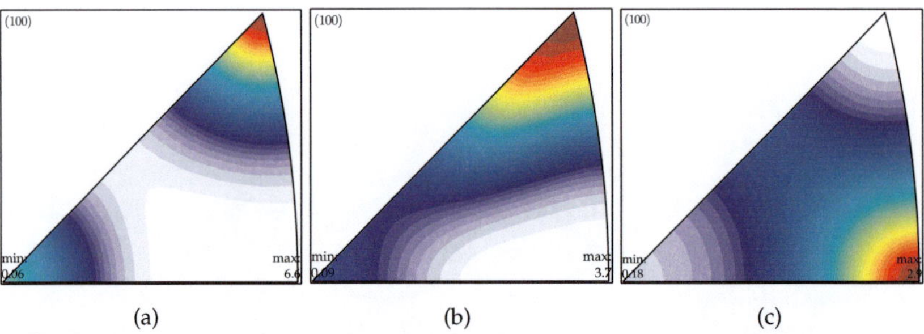

(a)   (b)   (c)

Figure 2.8: Representative triangle of continuous inverse pole figures of (a) tensile, (b) rolling and (c) shear texture

Euler space

An orientation can be easily represented in a three-dimensional space using the three parameters defining the orientation as coordinates. Therefore, the orientation space has several representations among which the Euler space is the one most frequently used (Hansen et al., 1978). Using a Cartesian coordinate system with the three Euler angles as coordinates, the rectangular Euler space shown in Fig. 2.9 is obtained. Since throughout this work, cubic symmetry will be used, in Fig. 2.9, the entire Euler space is already subdivided into 24 cubic fundamental zones. Each of the eight cuboids with dimensions 2π, $\pi/2$, $\pi/2$ in φ_1, Φ, φ_2-directions, respectively, which are the result of symmetry considerations rotating by π and $\pi/2$ about the (100)-axes, are, furthermore, decomposed into three equivalent curvilinearly bounded fundamental zones resulting from symmetry rotations about the (111)-axes.

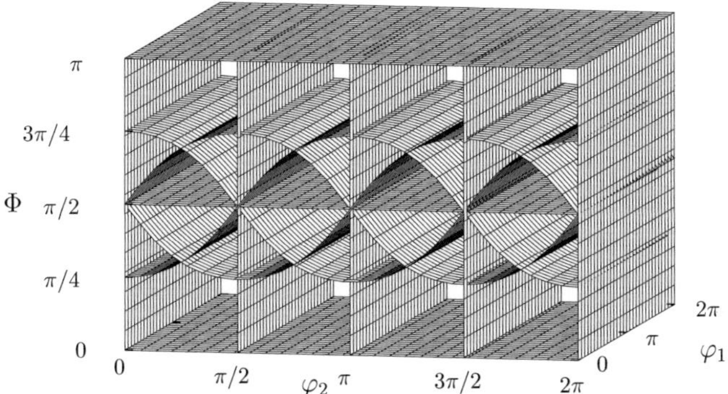

Figure 2.9: Cartesian Euler space with cubic fundamental zones

In case of an additional orthorhombic sample symmetry of the aggregate, the number of subdivisions of the orientation space becomes 96 $(= 24 \times 4)$ (cf. section 2.3.2). In this case, the φ_1-direction is, furthermore, quartered compared to the aforementioned case of triclinic sample symmetry. Attention has to be paid concerning the three equivalent fundamental zones in the cuboids, since the same orientation in case of a triclinic or an orthorhombic sample symmetry does not only differ in its φ_1-coordinate, but also changes its location in the Φ-φ_2-plane of the third fundamental zone (Fig. 2.10). This fact will be of special interest in section 4.1 for the partitioning procedure of the Euler space to reduce texture data.

To display orientation distributions of polycrystalline aggregates, usually a continuous representation within the Euler space is utilized. Therefore, similar to the procedure for pole figures, the application of particular functions allows to find a continuous approximation of the CODF. Then, sections of the Euler space are drawn showing contour lines of the three-dimensional function. It is common to plot φ_2-sections of the orientation space as shown in Fig. 2.11.

One has to be aware of the special case of $\Phi = 0$, where the Euler space is degenerated. Since for $\Phi = 0$, there is no uniqueness of the first and third Euler

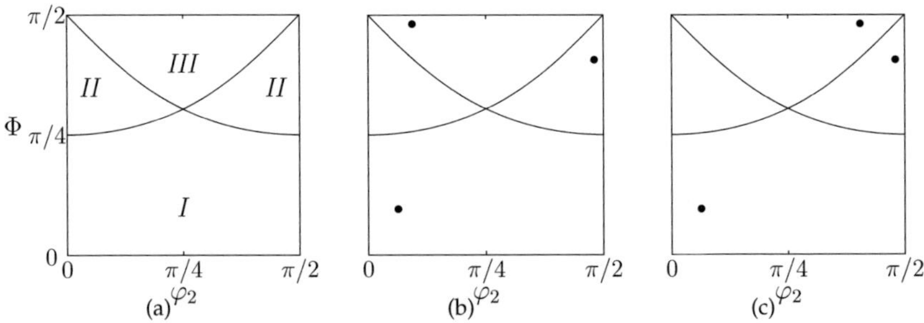

Figure 2.10:   Example of the same orientation in three equivalent cubic fundamental zones: (a) three types of fundamental zones, (b) orientation of cubic crystal with triclinic sample symmetry $(\varphi_1\,\Phi\,\varphi_2) =$ (0.6 0.3 0.2; 2.36 1.51 0.29; 3.95 1.28 1.51), (c) same orientation of cubic crystal with orthorhombic sample symmetry $(\varphi_1\,\Phi\,\varphi_2) =$ (0.6 0.3 0.2; 0.78 1.51 1.28; 0.81 1.28 1.51)

angles, but only for the sum of them (cf. Eq. 2.37), an orientation, which is usually represented by a point in the orientation space, is represented by a line in case of $\Phi = 0$. Inspired by Hansen et al. (1978), this degeneracy is depicted in Fig. 2.12 in terms of clouds of points having a certain misorientation with respect to three symmetrically equivalent ideal orientations. The ideal orientations are taken to be the same as in Fig. 2.10. Since these orientations are located near the borders of the depicted part of the orientation space, the clouds, corresponding to the three orientations, are cut at the borders extending into the Euler space at another location of the respective fundamental zone. Nevertheless, it becomes

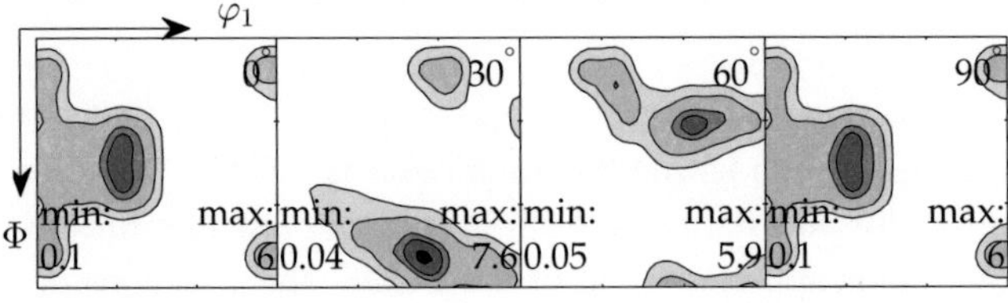

Figure 2.11: Representation of the orientation distribution by contour plots of the CODF in φ_2-sections of the Euler space using the example of an aluminum rolling texture

obvious that spherical clouds become ellipsoids for decreasing values Φ of the ideal orientations. This distortion of the Euler space should also be kept in mind for the texture analysis near this singularity.

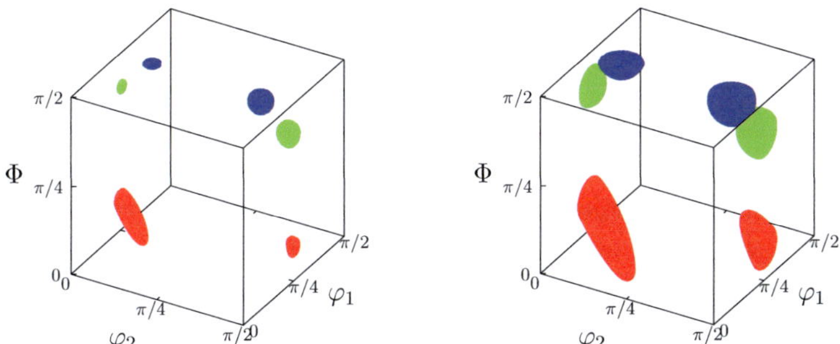

Figure 2.12: Degeneracy of the Euler space visualized by three symmetrically equivalent orientations and all points with a misorientation with respect to the ideal orientations smaller than $5°$ (left) and $10°$ (right)

2.3.5 Determination of mean orientations using quaternions

In Chapter 4, two methods will be introduced by which large orientation data sets are reduced. Both methods use different approaches to determine a specific number of similar discrete orientations but jointly necessitate a procedure to combine this set to one single representative. This section aims to discuss possibilities on finding a mean orientation of a set of orientations, which are then applied in sections 4.1 and 4.2.

Humbert et al. (1996) presented two methods on determining the mean orientation of a set of orientations involving quaternions. A quaternion $q = q(q_0, q_1, q_2, q_3)$ is associated to an orientation Q (Altmann, 1986; Becker and Panchanadeeswaran, 1989). The components $q_0, \dots, q_3$ can be computed, e.g., from Euler angles or a rotation angle and axis.

As Morawiec (1998b) discussed, the two methods of Humbert et al. (1996) do have shortcomings. The widely used method on determining the mean orientation q^m as the weighted arithmetic mean of the N single quaternions

$$q^m = \frac{1}{w} \sum_{i=1}^{N} w_i q_i,$$ (2.43)

with w_i, the weight of the ith quaternion q_i has the drawback that the determination of the quaternion q from the orientation Q is not unique, since

$$Q \rightarrow \pm q.$$ (2.44)

Although Humbert et al. (1998) responded that Eq. (2.43) can be anyway used since q_0 is always non-negative if the angle of rotation in the axis-angle convention is defined in the range $\omega \in [-\pi, \pi]$ and, therefore, $q_0 = \cos(\omega/2) \geq 0$, Morawiec (1998a) (see also Morawiec (2004)) proposed to use the second method of Humbert et al. (1996), however, with quaternions instead of orthogonal matrices to overcome difficulties according to improper rotations. The method is based on a minimization of the sum of all distances of the set of orientations to the sought mean orientation. One possibility to measure a distance between orientations is the misorientation which can be measured in terms of a rotation angle ω_{mis} related to a rotation ΔQ, by which one orientation Q_2 is transformed into another one Q_1

$$\cos \omega_{\text{mis}} = \frac{1}{2}(\text{tr}(\Delta Q) - 1), \quad \Delta Q = Q_1 Q_2^\mathsf{T}.$$ (2.45)

Morawiec, however, decided (according to Humbert et al. (1996)) to use another measure as a distance, being a Euclidean distance

$$d^2 = \frac{\|Q_1 - Q_2\|}{2} = 3 - \text{tr}(\Delta Q),$$ (2.46)

which, nevertheless, can be related to the misorientation angle

$$d^2 = 3 - \text{tr}(\Delta Q) = 2 - 2 \cos \omega_{\text{mis}} \stackrel{\text{small angles}}{\approx} \omega_{\text{mis}}^2.$$ (2.47)

The function to be minimized, thus, is taken to be

$$f(Q^M) = \sum_{i=1}^{N} w_i(d_i^2 - 3) = - \sum_{i=1}^{N} w_i \text{tr}(\Delta Q_i), \quad \text{tr}(\Delta Q_i) = Q^M \cdot Q_i,$$ (2.48)

where Q^M is the unknown mean orientation. Formulating Eq. (2.48) in terms of quaternions by accounting for the transformation

$$Q = (q_0^2 - V \cdot V)I + 2V \otimes V - q_0 \epsilon[V], \quad V = q_i e_i,$$ (2.49)

the function to be minimized reads

$$\tilde{f}(\boldsymbol{q}^M) = -\sum_{i=1}^{N} w_i(4(\boldsymbol{q}_i \cdot \boldsymbol{q}^M)^2 - \boldsymbol{q}^M \cdot \boldsymbol{q}^M), \tag{2.50}$$

with the unknown mean quaternion $\boldsymbol{q}^M$ and the dot product for quaternions being $\boldsymbol{q} \cdot \boldsymbol{q} = \sum_{k=0}^{3} q_k^2$. Since $\boldsymbol{q}^M$ needs to fulfill $\|\boldsymbol{q}^M\| = 1$, this side condition can be enforced introducing a Lagrangian multiplier λ, so that

$$\begin{aligned}
\hat{f}(\boldsymbol{q}^M) &= -\sum_{i=1}^{N} w_i(4(\boldsymbol{q}_i \cdot \boldsymbol{q}^M)^2 - \boldsymbol{q}^M \cdot \boldsymbol{q}^M) + \lambda(\boldsymbol{q}^M \cdot \boldsymbol{q}^M - 1) \\
&= \underbrace{-\sum_{i=1}^{N} w_i(4\boldsymbol{q}_i \otimes \boldsymbol{q}_i - \boldsymbol{I})}_{\boldsymbol{T}} \cdot (\boldsymbol{q}^M \otimes \boldsymbol{q}^M) + \lambda(\boldsymbol{q}^M \cdot \boldsymbol{q}^M - 1)
\end{aligned} \tag{2.51}$$

has to be solved for $\boldsymbol{q}^M$. Note that in this subsection, $\boldsymbol{I}$ is interpreted as a unit tensor in $\mathcal{R}^4$.

Demanding stationarity of Eq. (2.51) leads to an eigenvalue problem

$$\frac{\partial \hat{f}(\boldsymbol{q}^M)}{\partial \boldsymbol{q}^M} \stackrel{!}{=} \boldsymbol{o} \rightarrow \boldsymbol{T}\boldsymbol{q}^M = \lambda \boldsymbol{q}^M \quad \text{and} \quad \tilde{f}(\boldsymbol{q}^M) = -\lambda. \tag{2.52}$$

Therefore, it can be concluded that $\tilde{f}(\boldsymbol{q}^M)$ is minimal for $\boldsymbol{q}^M$ being the eigenvector corresponding to the largest eigenvalue of $\boldsymbol{T}$.[2]

2.4 General theory of homogenization

Since real materials possess a heterogeneous microstructure which significantly influences the macroscopic behavior, it is important to accurately examine the micro conditions and relations to properly describe the material behavior on the macro level. Therefore, using micromechanics, it is essential to separate the effects on the micro and macro scales (depending on the investigated material, it can be recommendable to even introduce a meso scale) and to investigate the mechanical relation between these effects. Fig. 2.13 shows different scales and the corresponding characteristic lengths for the example of a metal component.

In the sequel, some basics concerning scale transition are given, whereas in Chapters 5 and 6 particular homogenization methods are discussed in detail.

---

[2]This method is implemented in MTEX using the command mean(class quaternion).

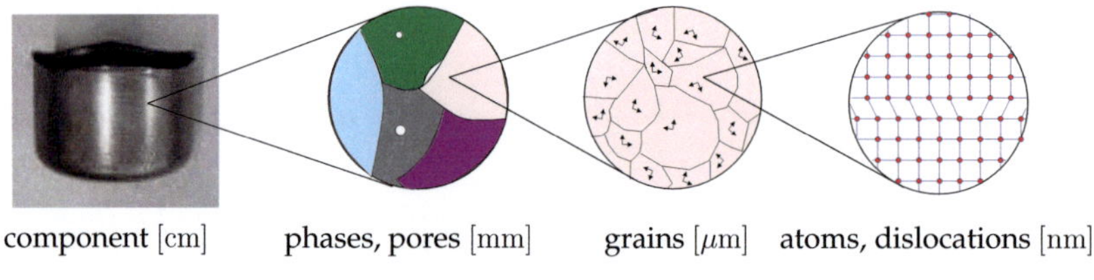

Figure 2.13: Characteristic scales [3]

For some more aspects of micromechanics and homogenization, the reader should refer to literature like Mura (1987); Torquato (2002); Gross and Seelig (2007); Ostoja-Starzewski (2008).

2.4.1 Scale transition and effective quantities

In this work, microscopically heterogeneous materials are considered, where the size of the heterogeneities (inclusions, inhomogeneities, grains, etc.) is small compared to the size of the specimen, so that the specimen can be treated as macroscopically homogeneous. Assuming additionally that the fluctuations of the mechanical fields on the micro scale are much smaller than the macro field fluctuations, the heterogeneous microstructural conditions can be averaged to describe the macroscopically homogeneous material behavior in terms of effective quantities (Hill, 1963, 1967). This procedure is called *homogenization*, whereas the inverse description estimating microscopic behavior, starting from the knowledge of macroscopic quantities (as, e.g., macroscopically observable stresses and strains), is called *localization* (see. Fig. 2.14). Hill (1963) proposed to introduce a representative volume element (RVE) at each material point of the macroscopic specimen, the size of which at the same time needs to be sufficiently large to be representative for the microstructure of the specimen, and sufficiently small compared to the overall size of the specimen. Then, the mechanical behavior of the RVE is assumed to represent the conditions of the entire specimen. Using such a RVE, the effective macroscopic mechanical fields of stresses and strains are defined as volume averages of the fluctuating micro

---

[3]Picture of drawn cup taken from experiments at the Institute for Metal Forming Technology, University of Stuttgart, that have been carried out within the scope of the Research Training Group 1483.

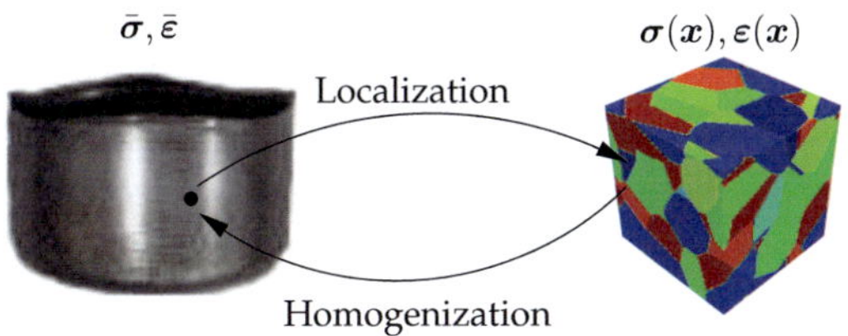

Figure 2.14: Localization and homogenization procedure

fields

$$\bar{\sigma} := \langle \sigma \rangle = \frac{1}{V} \int_{\text{RVE}} \sigma(x)\, dV, \qquad \bar{\varepsilon} := \langle \varepsilon \rangle = \frac{1}{V} \int_{\text{RVE}} \varepsilon(x)\, dV. \qquad (2.53)$$

The overbar $(\bar{\cdot})$ denotes effective quantities, and angular brackets $\langle \cdot \rangle$ denote the volume average (see also section 3.1). Using Gauss' theorem to transform volume into surface integrals, macroscopic stresses and strains can be equivalently determined using the traction t and displacement u on the boundary ∂V of the RVE

$$\bar{\sigma} = \frac{1}{V} \int_{\partial V} t(x) \otimes x\, dA, \quad \bar{\varepsilon} = \frac{1}{2V} \int_{\partial V} (u(x) \otimes n(x) + n(x) \otimes u(x))\, dA, \quad (2.54)$$

with n the outer normal of the boundary of the RVE. Although differentiable fields σ and u are required to apply Gauss' theorem, which is not fulfilled in case of heterogeneous materials with discontinuous material behavior, Eq. (2.54) even holds in the case of cracks or voids (see, e.g., Gross and Seelig (2007)).

Since the total volume V can be subdivided into perfectly bonded discrete subvolumes of the phases possessing homogeneous material behaviors, similarly, phase averages are defined as volume averages of the respective phase α covering the volume V_α

$$\langle \sigma \rangle_\alpha = \frac{1}{V_\alpha} \int_{V_\alpha} \sigma(x)\, dV, \qquad \langle \varepsilon \rangle_\alpha = \frac{1}{V_\alpha} \int_{V_\alpha} \varepsilon(x)\, dV. \qquad (2.55)$$

Denoting the concentration or volume fraction of phase α by $c_\alpha = V_\alpha/V$, Eq. 2.53 reads

$$\bar{\sigma} = \sum_{\alpha=1}^{N} c_\alpha \langle \sigma \rangle_\alpha, \qquad \bar{\varepsilon} = \sum_{\alpha=1}^{N} c_\alpha \langle \varepsilon \rangle_\alpha. \qquad (2.56)$$

2.4.2 Hill condition

The question of computing effective stress and strain fields, $\bar{\sigma}$ and $\bar{\varepsilon}$, on the basis of micro fields $\sigma(x)$ and $\varepsilon(x)$, respectively, yielded a fundamental result for the homogenization theory. Hill (1963) stated that the macroscopic strain energy and the volume average of the microscopic strain energy should be equivalent, so that

$$\langle \sigma \cdot \varepsilon \rangle = \langle \sigma \rangle \cdot \langle \varepsilon \rangle = \bar{\sigma} \cdot \bar{\varepsilon}. \tag{2.57}$$

Simultaneously, when splitting the stress and strain fields into a mean and fluctuating part

$$\sigma(x) = \langle \sigma \rangle + \tilde{\sigma}(x), \tag{2.58}$$
$$\varepsilon(x) = \langle \varepsilon \rangle + \tilde{\varepsilon}(x), \tag{2.59}$$

the Hill condition in Eq. (2.57) implies

$$\langle \tilde{\sigma}(x) \cdot \tilde{\varepsilon}(x) \rangle = 0. \tag{2.60}$$

The interpretation of the latter form of the Hill condition is that in average no physical work is done by the stress fluctuations on the strain fluctuations, or that in average, the stress and strain fluctuations are orthogonal.

It should be noted that the only requirements for the fields $\sigma(x)$ and $\varepsilon(x)$ are the equilibrium of the stress field $\mathrm{div}\,(\sigma)\,(x) = o$ and the compatibility of the strain field $\varepsilon(x) = \mathrm{sym}(\mathrm{grad}\,(u(x)))$, which means that even a constitutive coupling of $\sigma(x)$ and $\varepsilon(x)$ is not necessary. Therefore, these fields can in general appertain to different boundary value problems.

The above stated Hill condition is valid for

- homogeneous Neumann boundary conditions $t(x) = \langle \sigma \rangle n(x)$, $\forall x \in \partial V$,
- linear Dirichlet boundary conditions $u(x) = \langle \varepsilon \rangle x$, $\forall x \in \partial V$,
- periodic boundary conditions (on opposite boundaries ∂V^+ and ∂V^- of the RVE, the displacement field is periodic $u^+ = u^-$ and the tractions are antiperiodic $t^+ = -t^-$),
- $V \to \infty$, in case the ergodicity hypothesis (cf. section 3.1.2) holds.

Chapter 3

Microstructure of Polycrystalline Materials

3.1 Mathematical description of microstructures

3.1.1 Correlation functions

A heterogeneous medium is understood to be a material, which is composed of domains of different materials, as, e.g., in composites or of the same material in different states, e.g., in terms of different grains of polycrystals (Torquato, 2002). In general, the microstructure of such a medium, therefore, is described to consist of N discrete phases. The realization of each sample of the medium is the resultant of a stochastic process determining the particular microstructure (Torquato, 2002). Therefore, a mechanical variable ψ in statics depends on the location in space and on the particular realization of the random production process $\psi = \psi(\boldsymbol{x}, \omega)$, with $\omega \in \Omega$ and Ω, the set of all possible realizations called ensemble. To eliminate the influence of the random process on the mechanical variable, the ensemble average $\langle \cdot \rangle_\Omega$ is defined

$$\langle \psi \rangle_\Omega(\boldsymbol{x}) = \lim_{M \to \infty} \frac{1}{M} \sum_{i=1}^{M} \psi(\boldsymbol{x}, \omega_i). \tag{3.1}$$

Furthermore, a phase indicator function $\mathcal{I}_\alpha$ is introduced

$$\mathcal{I}_\alpha(\boldsymbol{x}, \omega) = \begin{cases} 1 & \forall \boldsymbol{x} \in V_\alpha(\omega) \\ 0 & \text{otherwise} \end{cases}, \tag{3.2}$$

denoting in a certain realization ω whether a point with position vector $\boldsymbol{x}$ is located in phase α ($\alpha = 1, \ldots, N$) or not, where $\sum_{\alpha=1}^{N} V_\alpha(\omega) = V(\omega)$. The indicator function possesses typical characteristics of projectors, i.e.,

$$\sum_{\alpha=1}^{N} \mathcal{I}_\alpha(\boldsymbol{x}, \omega) = 1 \quad \text{completeness,}$$
$$\mathcal{I}_\alpha(\boldsymbol{x}, \omega)\mathcal{I}_\alpha(\boldsymbol{x}, \omega) = \mathcal{I}_\alpha(\boldsymbol{x}, \omega) \quad \text{idempotence,} \tag{3.3}$$
$$\mathcal{I}_\alpha(\boldsymbol{x}, \omega)\mathcal{I}_\beta(\boldsymbol{x}, \omega) = 0, \ \alpha \neq \beta \quad \text{biorthogonality.}$$

The one-point correlation function $S_1^\alpha(\boldsymbol{x})$ gives the probability of finding phase α at position $\boldsymbol{x}$ in the ensemble

$$S_1^\alpha(\boldsymbol{x}) = \langle \mathcal{I}_\alpha(\boldsymbol{x}, \omega) \rangle_\Omega. \tag{3.4}$$

$S_1^\alpha(\boldsymbol{x})$ can be interpreted as the position dependent volume fraction of phase α. Similar to Eq. (3.4), one defines higher order correlation functions (n-point correlation functions), giving the probability of finding the phase α simultaneously at n points $\boldsymbol{x}_1, \ldots, \boldsymbol{x}_n$

$$S_n^\alpha(\boldsymbol{x}_1, \boldsymbol{x}_2, \ldots, \boldsymbol{x}_n) = \langle \mathcal{I}_\alpha(\boldsymbol{x}_1, \omega) \mathcal{I}_\alpha(\boldsymbol{x}_2, \omega) \ldots \mathcal{I}_\alpha(\boldsymbol{x}_n, \omega) \rangle_\Omega. \tag{3.5}$$

Furthermore, mixed n-point correlation functions can be evaluated, as, e.g., $S_3^{\alpha\beta\gamma}(\boldsymbol{x}_1, \boldsymbol{x}_2, \boldsymbol{x}_3) = \langle \mathcal{I}_\alpha(\boldsymbol{x}_1, \omega) \mathcal{I}_\beta(\boldsymbol{x}_2, \omega) \mathcal{I}_\gamma(\boldsymbol{x}_3, \omega) \rangle_\Omega$, being the 3-point probability of finding simultaneously phase α at $\boldsymbol{x}_1$, phase β at $\boldsymbol{x}_2$ and phase γ at $\boldsymbol{x}_3$. The latter case of mixed probability functions, which is less important for this work, is, however, given for the sake of completeness. Certainly, besides the correlation functions, several other important statistical descriptors for microstructures exist, being not discussed in this work. For an extensive treatise of this subject, the reader is referred to the books of Ohser and Mücklich (2000) and Torquato (2002).

3.1.2 Ergodicity and statistical homogeneity

In the previous section it was shown that the correlation functions are position dependent in general, which is the case for statistically inhomogeneous materials. If the correlation functions are invariant under a translation $\Delta\boldsymbol{x}$, i.e.,

$$S_n^\alpha(\boldsymbol{x}_1, \boldsymbol{x}_2, \ldots, \boldsymbol{x}_n) = S_n^\alpha(\boldsymbol{x}_1 + \Delta\boldsymbol{x}, \boldsymbol{x}_2 + \Delta\boldsymbol{x}, \ldots, \boldsymbol{x}_n + \Delta\boldsymbol{x}) \tag{3.6}$$

statistically homogeneous media are described. In this case, the one-point correlation function does not depend on the particular position and, therefore, is the constant volume fraction $S_1^\alpha = c_\alpha$. However, the n-point correlation functions depend on the vectorial distance between the considered points but they are independent of the particular position. Furthermore, statistically homogeneous media are said to be statistically isotropic if they are rotationally invariant, which means they only depend on scalar distances between the points.

Being concerned with a statistically homogeneous medium where different spatial regions have similar properties, it is useful to relate volume and ensemble averages. The hypothesis of ergodicity states that averaging a mechanical field $\psi(\boldsymbol{x}, \omega)$ at a fixed point $\boldsymbol{x}_0$ over the ensemble is equivalent to averaging the same field over the volume of only one realization ω_0 among the ensemble, if the volume of the sample is infinitely large

$$\langle \psi(\boldsymbol{x}_0, \omega) \rangle_\Omega = \langle \psi(\boldsymbol{x}, \omega_0) \rangle \quad \text{if} \quad V \to \infty. \tag{3.7}$$

For these ergodic media, it is an advantage to compute the correlation functions using the volume average instead of ensemble averaging, since, usually, microstructural data is available for only one or tentatively a few realizations of the investigated sample.

3.2 Real microstructures

The electron backscatter diffraction (EBSD) method allows to characterize phases and grain orientations on a surface of a crystalline material (see, e.g., Schwartz et al. (2000); Engler and Randle (2009); Schwartz et al. (2009)). Using a scanning electron microscope (SEM) or a transmission electron microscope (TEM), the electron beam is focused on the specimen with an angle of usually $70°$ to the horizontal. The diffraction pattern resulting from the interaction of the electron beam with the atoms of the material is detected on a phosphor screen. Due to interference (Bragg condition), bright bands, i.e., the particular Kikuchi pattern (Kikuchi, 1928), become observable representing the crystal orientation including the crystal symmetry. The experimental setup of an EBSD arrangement and, exemplarily, a Kikuchi or electron backscatter pattern (EBSP) are shown in Fig. 3.1.

The Kikuchi pattern is detected at discrete points in a regular grid on the specimen surface. However, the facilitative automatic indexing, i.e., computation of crystal orientations by means of automatic recognition of Kikuchi bands, was not developed until the 1990s (Schwartz et al., 2000). As output, the EBSD measurement mainly provides information about 1.) the phase, 2.) the location of the measurement point, 3.) the grain orientation, 4.) the reliability of the result in terms of a mean angle (MAD), characterizing the deviation of the measured and calculated pattern, and 5.) the number of bands of the pattern used for

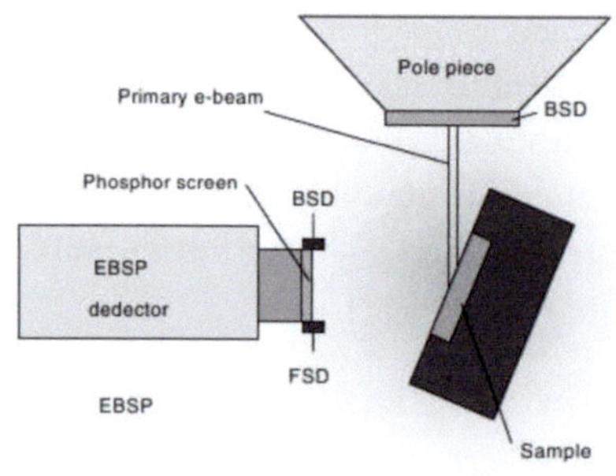

Figure 3.1: Sketch of an EBSD measurement (left; Zhou and Wang (2007)) and example of a Kikuchi pattern (right; Schwartz et al. (2009))

obtaining the result. A microstructural image is usually directly shown by the software of the EBSD equipment using the inverse pole figure (see section 2.3.4) for coloring different grains according to their crystal orientation (Fig. 3.2)

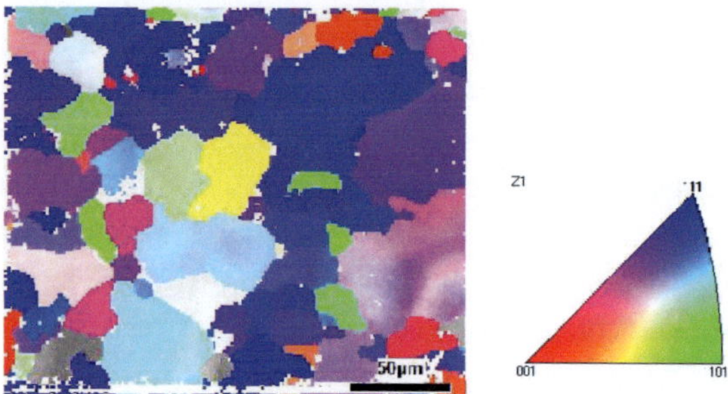

Figure 3.2: Microstructural image obtained by EBSD measurement with inverse pole figure coloring (Project A6, GRK 1483)

By the EBSD technique, not only two-dimensional microstructural information is detected. Moreover, by cutting sections of the specimen, e.g., using a focused ion beam (FIB), the three-dimensional data of the microstructure can be measured and reconstructed. Since the latter method is destructive, three-dimensional EBSD data can not be recorded to continuously track the texture evolution, in contrast to 2D EBSD measurements, which can be carried out almost in-situ during deformation experiments (usually in a loop of unloading, followed by an EBSD measurement, subsequent reloading, etc.).

The two- or three-dimensional EBSD images can be directly discretized to be computable by the Finite Element Method (FEM), e.g., by the Simpleware[4] software. Although these microstructure-based FE models are usually pretty huge due to the meshing of the complex geometry of, e.g., grain boundaries of real microstructures, this procedure is very useful when investigating the mechanical behavior of a particular or representative realization of the produced specimen. Since this work is addressed to fundamental investigations of effects of crystallographic and morphological texture on the mechanical response of polycrystals, mainly artificial microstructures are used for spatially resolved computations. This approach facilitates the generation of plenty of different microstructures to draw statistical conclusions from various computations. Nevertheless, the EBSD data is important in this work to provide real crystallographic textures to the simulations and to assure the proper statistics of the artificial microstructures.

3.3 Artificial microstructures of Voronoi type

In the literature, it is still popular to generate cuboidal grains in order to simplify the computational models, especially to produce reference solutions for semi-analytical homogenization methods (e.g., Ma, 2003; Tadano et al., 2012), to study size influences of RVEs (e.g., Ranganathan and Ostoja-Starzewski, 2008; Salahouelhadj and Haddadi, 2010), or in virtual material testing to evaluate yield loci (e.g., Kraska et al., 2009; Kim et al., 2012). Delannay et al. (2006) investigated cuboidal and truncated octahedron grains in FE polycrystal models showing that the latter improves results for strain field distribution and texture evolution compared to experiments. Throughout this work, artificial microstructures are generated based on Voronoi mosaics.

3.3.1 Classical Voronoi tesselation

When analyzing real microstructures, it becomes obvious that the individual grains of polycrystalline materials nearly possess polyhedral shape. This fact

---

[4]http://www.simpleware.com

is the result of the formation process, where crystals grow out of the cast. If a homogeneous cooling rate is present throughout the cast, and crystals start to grow from randomly distributed points at the same time in an isotropic manner up to they contact with a neighboring grain, at the bisector of the distance between the two initial crystal nuclei, a grain boundary forms. Of course, this is the very idealized understanding of crystal growth. In reality, there appear other effects such as grain boundary formation due to energy minimization, which distract the crystal from being such a polyhedron. Furthermore, it has been found that the grain size fluctuation and the number of neighbors of the grains does not satisfactorily comply with the situation in real materials (Döbrich et al., 2004). Nevertheless, the method of forming a Voronoi mosaic is widely accepted to create artificial polycrystalline microstructures (Ohser and Mücklich, 2000). Fig. 3.3 depicts the construction scheme for a classical in plane Voronoi tessellation with five nuclei. Basic properties of Voronoi mosaics as well as special purpose Voronoi tessellations modifying the initial point seed and generalizations, e.g., Johnson-Mehl and Laguerre tesselations, are discussed in Aurenhammer (1991); Ohser and Mücklich (2000); Ostoja-Starzewski (2008); Lautensack and Zuyev (2008).

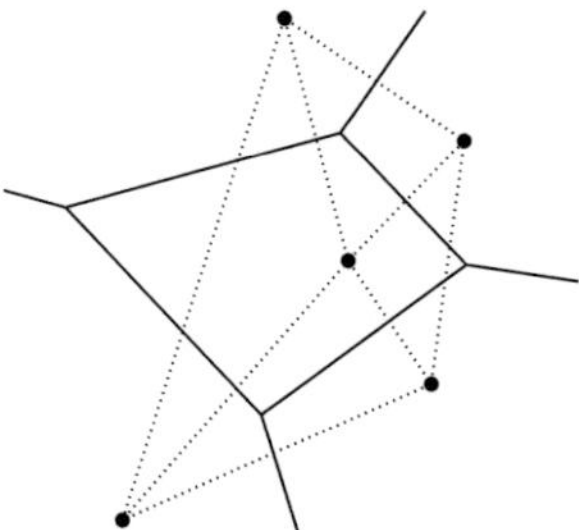

Figure 3.3: Construction of classical two-dimensional Voronoi mosaic with five generator points

Throughout this work, mainly the classical Poisson-Voronoi mosaic is used for representing the microstructure of single-phase polycrystals. Different strategies exist for generating computational models based on Voronoi-type microstructures. Among others, Musienko and Cailletaud (2009); Böhlke et al. (2009); Jöchen et al. (2010); Fritzen and Böhlke (2011); Quey et al. (2011) used the classical and special-purpose Voronoi mosaics in order to model polycrystalline aggregates. In these cases, the finite element mesh was created after fixing the geometry, in order to exactly resolve grain boundaries with the mesh. In contrast, the present work uses structured meshes overlying the microstructural geometry. This is done by first generating the mesh of the specimen independent of the underlying microstructure. Then, the microstructure is generated by determining which finite elements or integration points of the finite elements belong to which grain. This approach yields cascaded grain boundaries or multi-grain elements, so that the grain boundaries do not coincide with element boundaries (Fig. 3.4(a)). By this procedure, the same FE mesh can be used in various simulations with different microstructures to draw statistical conclusions, reducing the entire computational effort enormously. Although grain boundaries and, therefore, also stress and strain fields are not resolved accurately by using the structured-mesh-approach, Böhlke et al. (2010) have shown that, nevertheless, the statistics of stress and strain fields are approximately identical in both cases (see also section 5.7).

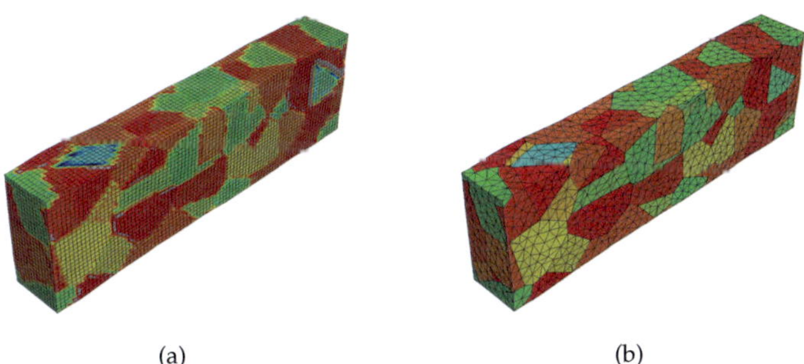

(a)   (b)

Figure 3.4: (a) Structured and (b) unstructured mesh of Voronoi-type microstructure (Böhlke et al., 2010)

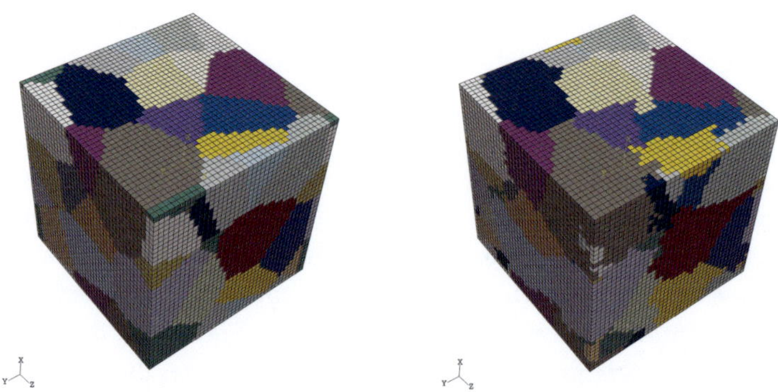

Figure 3.5: Initial (left) and converged (right) periodic microstructure with 25 grains

3.3.2 Voronoi-type mosaic with prescribed volume fractions

For the purpose of comparison with semi-analytical methods, it is required to generate a microstructural FE model with the same volume fraction distribution as used in the semi-analytical scheme. Especially when a technique to reduce the data of a measured orientation distribution is applied (as, e.g., described in Chapter 4), that afterwards should enter a homogenization scheme, the resulting data contains a small set of mean orientations assigned to specific grain sizes. Therefore, this section aims to describe a method to shrink or grow the cells of a classical Voronoi mosaic in conjunction with the structured-mesh approach in order to obtain the required volume fraction distribution. In the first step, a classical periodic Voronoi tessellation is generated on a structured finite element mesh using a randomly distributed point seed for the required number of grains. The volume fractions of these Voronoi cells are assigned to the target volume fractions according to the grain size, so that the randomly generated largest grain shall shrink or grow to become the largest of the target grains and so forth. For each finite element that belongs to a border of a Voronoi cell, it is decided whether its cell should grow or shrink, and also for its neighbor of the other cell, so as to assign the element itself or the neighbor to the one or the

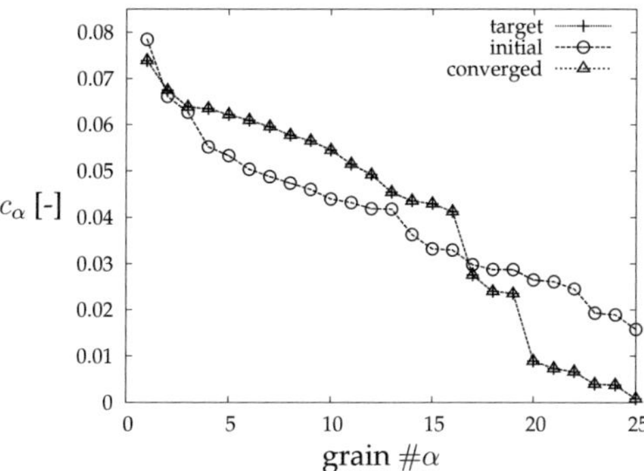

Figure 3.6: Initial, target and converged volume fraction distribution for 25 Voronoi cells

other grain. Three different cases are distinguished when examining a border element of one grain cell

- The cell of the border element shall grow, the neighboring cell shall shrink ⇒ Neighboring element is added to the border element's cell.
- The cell of the border element shall shrink, the neighboring cell shall grow ⇒ Border element is added to the neighboring element's cell.
- Both, the border and the neighboring element's cells should either shrink or grow ⇒ Nothing is changed.

This step is repeated for all border elements sequentially, up to the required volume fractions are reached. Fig. 3.5 shows a FE model consisting of 25 grains with the initial random Voronoi mosaic and the converged solution fulfilling the volume fraction distribution requirements. Additionally, Fig. 3.6 depicts the initial, reached and target volume fraction distributions for the example with the 25 grains. The generated microstructure shows to yield non-convex grains. If necessary, this effect could be diminished by, e.g., randomly selecting the sequence of border elements to request the need for shrinking or growing, by using a Hardcore-Voronoi tessellation as initial configuration or by applying a Monte-Carlo-simulation to increase the number of element neighbors of the same phase. Since visually, the obtained microstructure does not seem to be unrealistic, the procedure is identically applied as described above for the usage in the present work without any post-processing.

Chapter 4

Representative Reduction of Orientation Data

The description of the constitutive behavior of materials with microstructure can be realized in two different ways. On one hand, the microstructure can be accurately resolved by, e.g., discretizing microstructural images (from electron backscatter diffraction (EBSD), atomic force microscopy (AFM), computer tomography (CT), etc.; cf. Chapter 3) with finite elements. Due to the lack of computational power, by this technique, the boundary value problem of only a small part of a macroscopic structure can be solved efficiently. Usually, a representative fraction of the microstructure, i.e., a representative volume element (RVE), is considered. A further approach is to use homogenization methods (e.g., Mura, 1987; Qu and Cherkaoui, 2006; Böhlke et al., 2010)) treating the phasewise constitutive behavior in a mean sense (phase averages of stresses and strains). This allows for modeling the response of a whole microstructured component. For efficiently taking into account the underlying microstructure in these macroscopic simulations, it is necessary to identify main characteristics and, therefore, generating a low-dimensional description of the microstructure. Considering polycrystalline behavior, the orientation information, which is usually available at a certain measurement grid, should be reduced to main texture components (e.g., Kocks et al. (1998)) to enter the simulation framework. In this section, a method is described which allows to reduce orientation data of a microstructure to a variable but sufficiently small number of texture components.

4.1 Division of the Euler space

In order to identify main texture components from texture measurements, the orientation data is mapped into one fundamental zone (FZ) of the Euler space being appropriate for the present symmetries (cf. section 2.3.4). Following the work of Gao et al. (2006), who used a partitioning technique of the orientation space to compute correlation functions, one of the FZs is divided into a discrete number of boxes to reduce texture data.

In Jöchen and Böhlke (2011), this partitioning method is applied using the same FZ being valid for cubic crystal symmetry as in Gao et al. (2006) (Fig. 4.1(a)). This

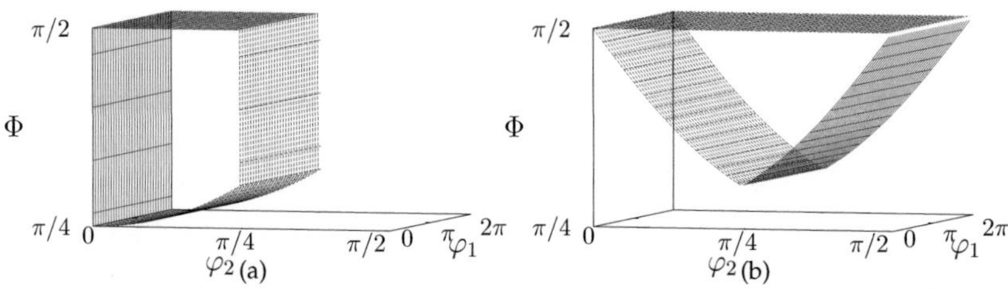

Figure 4.1: Cubic fundamental zones (a) chosen by Gao et al. (2006) and used for data reduction by Jöchen and Böhlke (2011) and (b) used in this thesis and in Jöchen and Böhlke (2012) for data reduction

fundamental zone does not coincide with the ones drawn in Fig. 2.10, but is a combination of parts of FZs *II* and *III*. The advantage of this choice of FZ is its nearly cuboidal shape with only one weakly non-linear boundary, simplifying significantly the partitioning into cuboidal boxes. For polycrystals with cubic crystal symmetry and triclinic sample symmetry, this FZ can be used just as well as one of the 24 FZs shown in Fig. 2.9. However, when investigating polycrystals with cubic crystal and orthotropic sample symmetry, only the FZs bounded as shown in Fig. 2.9 are admissible to be chosen.

For the two different cases of triclinic and orthotropic sample symmetry to be investigated in this work, the limits of the FZs in the $\Phi - \varphi_2$-plane are identical while only the limit in φ_1-direction differs, being given by

$$\varphi_1 \in \begin{cases} [0, 2\pi) & \text{triclinic sample symmetry} \\ \left[0, \dfrac{\pi}{2}\right) & \text{orthotropic sample symmetry} \end{cases} . \tag{4.1}$$

Nevertheless, the additional symmetry relations due to the sample symmetry do not only affect the φ_1-value of an orientation triple, but can as well change the φ_2-position of an orientation in the Euler space. This fact can be easily reproduced by examining the three symmetrically equivalent orientations depicted in Fig. 2.10 for the two distinguished cases. Therefore, the FZ chosen by Gao et al. (2006) is not appropriate for both cases of examined sample symmetries, so that in the sequel the FZ denoted with *III* in Fig. 2.10 is chosen for the data reduction technique.

This FZ *III*, being also depicted in Fig. 4.1(b), is delimited by the boundaries

$$0 \leq \varphi_1 < \begin{cases} 2\pi & \text{triclinic sample symmetry} \\ \dfrac{\pi}{2} & \text{orthotropic sample symmetry} \end{cases}$$

$$\arccos\left(\frac{\cos(\varphi_2 + 3\pi/2)}{\sqrt{1 + \cos^2(\varphi_2 + 3\pi/2)}}\right) \leq \Phi \leq \frac{\pi}{2} \qquad \forall \quad \varphi_2 \in \left[0, \frac{\pi}{4}\right], \tag{4.2}$$

$$\arccos\left(\frac{\cos(\varphi_2)}{\sqrt{1 + \cos^2(\varphi_2)}}\right) < \Phi < \frac{\pi}{2} \qquad \forall \quad \varphi_2 \in \left(\frac{\pi}{4}, \frac{\pi}{2}\right),$$

$$0 \leq \varphi_2 < \frac{\pi}{2}.$$

4.1.1 Rules of dissection

Constituting that the FZ is divided in N boxes, each of them occupying the equal volume fraction with respect to the volume of the complete orientation space

$$V(B_i) = \frac{1}{8\pi^2} \int\limits_{Box} \sin\Phi \, d\varphi_1 \, d\Phi \, d\varphi_2 = \frac{1}{F_{sym}N}, \tag{4.3}$$

$$F_{sym} = \begin{cases} 24 & \text{triclinic sample symmetry} \\ 96 & \text{orthotropic sample symmetry} \end{cases}, \tag{4.4}$$

the boundaries of each box can be computed iteratively. When denoting the number of boxes in φ_1-, Φ- and φ_2-direction by $i = 1,\ldots,I$, $j = 1,\ldots,J$ and $k = 1,\ldots,K$, respectively, so that $N = IJK$, the boundaries of the I boxes in φ_1-direction can be determined by

$$\varphi_1^i = \frac{\varphi_{1,max}}{I} i, \qquad \varphi_{1,max} = \frac{\pi}{2} \text{ or } 2\pi. \tag{4.5}$$

For the identification of the bounds of the boxes in the $\Phi - \varphi_2-$plane, the nonlinear limits of the fundamental zone (cf. Eq. 4.3) have to be accounted for. Therefore, starting with the volume of the whole cubic-triclinic FZ

$$V_{FZ} = \int\limits_{0}^{\pi/2} \int\limits_{\Phi_{NL}(\varphi_2)}^{\pi/2} \int\limits_{0}^{2\pi} \sin\Phi \, d\varphi_1 \, d\Phi \, d\varphi_2 = \frac{\pi^2}{3}, \tag{4.6}$$

with the non-linear boundary Φ_{NL} being given by Eq. (4.3)$_{2,3}$, the iterative determination of the limits of the boxes in φ_2-direction is based on the partitioning of the orientation space into equal volume slices

$$\int\limits_{\varphi_2^{k-1}}^{\varphi_2^k} \int\limits_{\Phi_{NL}(\varphi_2)}^{\pi/2} \int\limits_0^{2\pi} \sin\Phi \, d\varphi_1 \, d\Phi \, d\varphi_2 = \frac{\pi^2}{3K}. \tag{4.7}$$

Due to the symmetry of the FZ with respect to $\varphi_2 = \pi/4$, the boundaries in φ_2-direction for $k \in [0, K/2]$ which are to be determined solving

$$\int\limits_{\varphi_2^{k-1}}^{\varphi_2^k} \frac{\cos(\varphi_2 + 3\pi/2)}{\sqrt{1 + \cos^2(\varphi_2 + 3\pi/2)}} \, d\varphi_2 = \frac{\pi}{6K} \quad \text{with} \quad \varphi_2^0 = 0, \tag{4.8}$$

can be mirrored for the second half of the FZ. This implies K to be an even number. The boundaries of the boxes in Φ-direction can then be computed using Eq. (4.4) resulting in

$$\cos\Phi^{j-1} - \cos\Phi^j = \frac{\pi}{6JK(\varphi_2^k - \varphi_2^{k-1})}, \tag{4.9}$$

which is the easiest to be solved recursively using $\Phi^J = \pi/2$. Attention has to be paid concerning the integration of Eq. (4.4) for boxes that intersect the non-linear boundary of the FZ, which can be avoided in the middle part of the FZ when using the recursive computation mentioned above. Nevertheless, these intersections most likely occur near $\varphi_2 = 0$ and $\varphi_2 = \pi/2$ (cf. Fig. 4.2(a)). In this case, Eq. (4.9) has to be modified accounting for the point of intersection and the (partial) non-linear boundary of the box.

As becomes obvious by Fig. 4.2(a), the boxes near $\varphi_2 = 0$ and $\varphi_2 = \pi/2$ become elongated due to the side condition of volumetric identity of all boxes and a fixed number of boxes in Φ-direction independent of the position φ_2. Another possibility of partitioning to enforce approximately the same box shapes is to introduce a variable number of boxes in Φ-direction while still demanding equal volumes of boxes. In this sense, the number of boxes at $\varphi_2 = \pi/4$ is chosen to be J, being reduced by one with each step (i.e., boxwidth) towards $\varphi_2 = 0$ and $\varphi_2 = \pi/2$. The choice of varying the number of boxes by one along φ_2 is obvious due to the slope of the non-linear boundary of the FZ being approximately ± 1. In the sequel, this method is addressed as method 2 ($M2$), while the previous

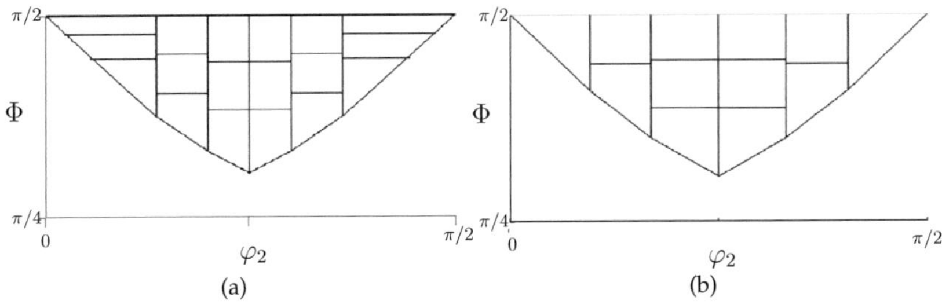

Figure 4.2: Example of partitioning of the cubic-triclinic fundamental zone in $\Phi - \varphi_2-$plane for $J = 3$ and $K = 6$: (a) fixed ($M1$) and (b) variable number of boxes in Φ-direction ($M2$)

one with a constant number of boxes in Φ-direction is called method 1 ($M1$). The total number N of boxes for this partitioning approach is given by

$$N = 2I \sum_{j=J-K/2+1}^{J} j = I\left((J+1)^2 - (J - \frac{K}{2} + 1)^2 - \frac{K}{2}\right), \quad \forall J \geq \frac{K}{2}. \qquad (4.10)$$

The boundaries in φ_1-direction are unaffected compared to the previous approach and the limits in φ_2-direction of the slices of the FZ can be computed similar to Eq. (4.8), taking, however, into account, that the volume of each slice depends on the position φ_2^k, so that for $k = 1, \ldots, K/2$

$$\int_{\varphi_2^{k-1}}^{\varphi_2^k} \frac{\cos(\varphi_2 + 3\pi/2)}{\sqrt{1 + \cos^2(\varphi_2 + 3\pi/2)}} \, d\varphi_2 = \frac{\pi(J - K/2 + k)}{12 \sum_{j=J-K/2+1}^{J} j} \quad \text{with} \quad \varphi_2^0 = 0 \qquad (4.11)$$

holds. The computation of boundaries in Φ-direction is then

$$\cos \Phi^{j-1} - \cos \Phi^j = \frac{\pi}{12(\varphi_2^k - \varphi_2^{k-1}) \sum_{j=J-K/2+1}^{J} j}. \qquad (4.12)$$

An example of box partitioning using the latter method is depicted in Fig. 4.2(b). After partitioning, the mean orientation of all measured orientations is computed using quaternions. The set of these mean orientations represents a reduced but representative data set of the initial data, so that the parameter N determines the amount of compression of the data.

4.1.2 Size and shape of domains

In order to quantify the quality of the low-dimensional texture representation, the texture index (Bunge and Roberts, 1969) can be used. The texture index being defined as the integral of the square root of the CODF

$$TI = \int_{SO(3)} f(\mathbf{Q})^2 \, dQ \qquad (4.13)$$

gives an impression about the sharpness of the texture. Furthermore, the difference of the approximated crystallite orientation distribution function to the CODF of the raw data is computed by

$$E = \int_{SO(3)} |f_{app}(\mathbf{Q}) - f(\mathbf{Q})|^2 \, dQ. \qquad (4.14)$$

Possibilities to improve the approximation of a CODF are, e.g., i) coarsening or refining the number of boxes, ii) translating the fixed number of boxes in space or iii) changing the box aspect ratio.

In order to evaluate the presented method, two data sets are examined in the sequel concerning the quality of the approximation mainly focusing on the influence of i) and iii). The data sets are 1) a uniform orientation distribution $(\max(f(\mathbf{Q})) = 1.2)^5$ and 2) a cube dominated texture of recrystallized aluminum alloy AA3104 before rolling $(\max(f(\mathbf{Q})) \approx 11$; data taken from Delannay et al. (2002)), both sets initially containing 10000 discrete orientations (Fig. 4.3).

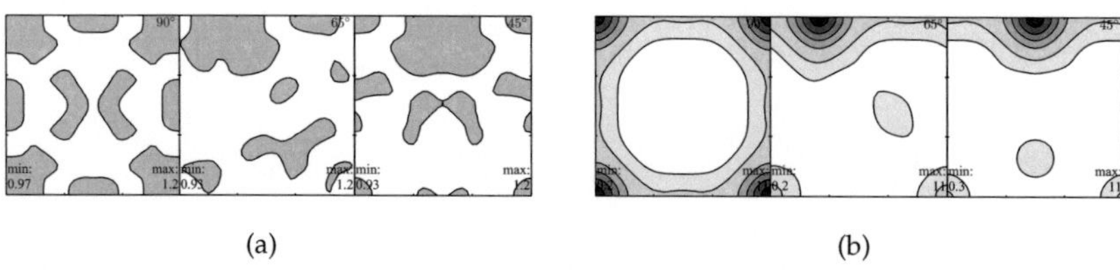

(a)   (b)

Figure 4.3: $\varphi_2 = (90°, 65°, 45°)$-slices of the CODF of a) gray texture b) recrystallized aluminum texture before rolling

---

[5]In this section, all CODFs have been evaluated by MTEX with a De-la-Vallee-Poussin kernel and a halfwidth of $7°$.

Influence of size of domains

In this section, the geometry of boxes is approximately kept constant, so that the total number of boxes in the three directions have a constant ratio of $I : J : K = 8 : 1 : 2$ according to the prismatic envelope of the cubic fundamental zone of the Euler space. Due to the nature of the second method, prescribing I, J and K leads to a smaller number of boxes for method 2 ($M2$) compared to the method 1 ($M1$), where the number of boxes in Φ-direction is also constant.

Figures 4.4 and 4.5 show the normalized texture index (texture index of approximation divided by texture index of the CODF of the raw data) as well as the error in the CODF with respect to the number of boxes (or texture components) for the two examined data sets. The results of all three proposed methods to determine

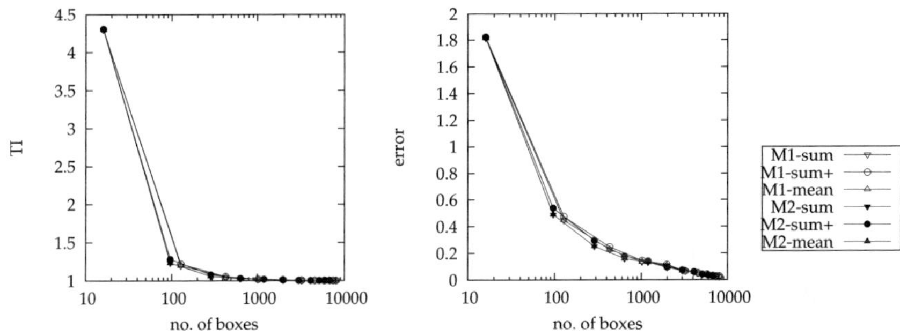

Figure 4.4: Normalized texture index (TI) and $L2$-error of the CODFs (error) for the uniform orientation distribution with respect to the number of boxes

the mean orientation (weighted sum of quaternions (sum), weighted sum of quaternions using only quaternions with $q_0 \geq 0$ (sum+) and the eigenvalue procedure (mean)) for the two methods ($M1$ and $M2$) of box partitioning are examined in these figures (cf. section 2.3.5). It becomes obvious that for both data sets, the normalized texture index is already smaller than TI $= 1,5$ for a box number of ≈ 100. However, for the aluminum texture, the normalized texture index increases again for box numbers in the range $100 - 400$, while for the gray texture, it strictly converges towards 1 for an increasing number of boxes. This can happen, since the textured data sets show clouds in the

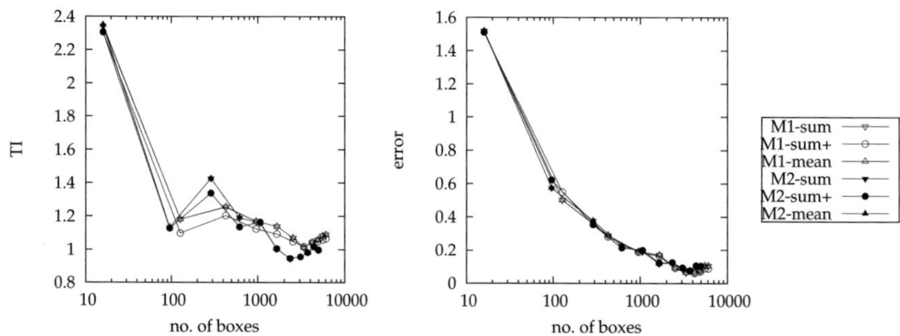

Figure 4.5: Normalized texture index (TI) and $L2$-error of the CODFs (error) for the aluminum texture with respect to the number of boxes

orientation space, which are either favorably or unfavorably divided by the partitioning technique in the sense of CODF approximation. Nevertheless, independent of the chosen data set, the figures show that in all cases the CODF error decreases with an increasing number of boxes. All averaging schemes and both partitioning methods yield approximately the same results in this study where the box geometry is chosen to have a constant aspect ratio. Remarkably, the methods sum and mean yield exactly the same results for these isotropic box morphologies.

Influence of domain geometry

To study the influence of the box shape on the CODF approximation, a fixed number of 300 boxes is chosen and examined. In contrast to usually utilized cubical-shaped boxes, the distribution of data points in the Euler space could necessitate different box morphologies to receive a good approximation, e.g., in case of elongated data point clouds. Poor approximations of the texture result in case of extremal values of box shapes, e.g., slice like boxes with only one of the Euler angle ranges being divided in more than one region (see Fig. 4.6).

Figures 4.7 and 4.8 show the distribution of CODF errors and normalized texture indices for both data sets and partitioning methods $M1$ and $M2$ and averaging techniques sum and mean. Although the method sum+ gives slightly different results compared to sum, the distributions look very similar, so that the results of sum+ are not depicted separately. In the aforementioned figures, black squares

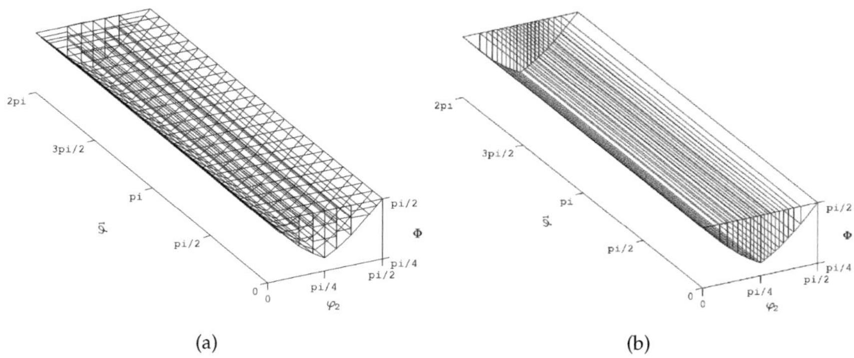

(a)   (b)

Figure 4.6: (a) Nearly isotropic and (b) anisotropic box aspect ratios generated with $M1$.

denote the minimum and white squares the maximum associated with the diagram. The numerical values are given in Table 4.1. The figures show all possibilities of generating 300 boxes with method $M1$ and $M2$, the latter offering only less variants due to the side condition $J \geq K/2$. The abscissa shows the number I and the ordinate the number J, while K is unique due to the total number of boxes being 300. It becomes obvious that the error and texture index distributions are similar for the methods sum and mean except for the case of $I = 1$. In this case, the methods generate very different results, being explainable by the very long boxes in φ_1-direction. Method mean should in general not be applicable for such a case, since the cloud of points is very much elongated so that ω_{mis} is no longer small. On the other hand, this method generates by chance a better approximation compared to sum, since the resulting mean orientations are much more distributed along φ_1, compared to the result of sum. For a better clarification, in Fig. 4.9 this is exemplarily depicted for the smaller number of 40 boxes but showing the same effect. For $I \geq 2$, however, this phenomenon vanishes. By summarizing the results of the figures, it can be concluded that for $I \geq 4$ the results improve and show the smallest errors and best agreements in the texture indices independent of the chosen data set.

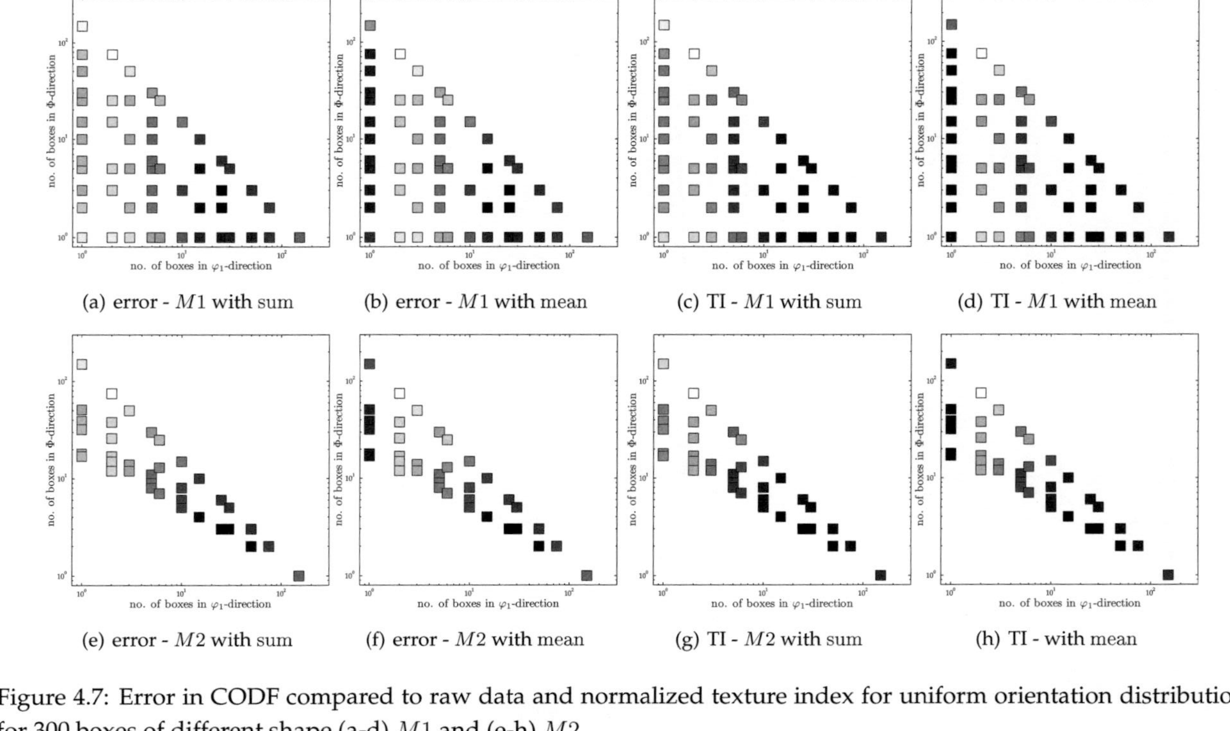

Figure 4.7: Error in CODF compared to raw data and normalized texture index for uniform orientation distribution for 300 boxes of different shape (a-d) $M1$ and (e-h) $M2$

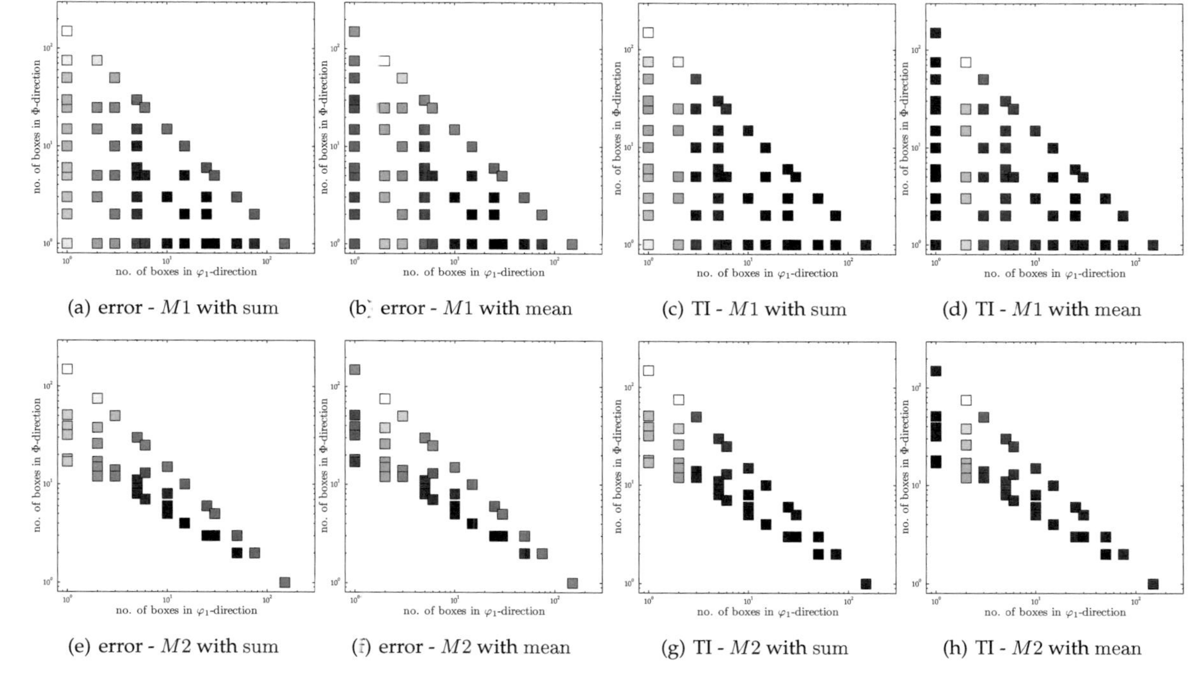

Figure 4.8: Error in CODF compared to raw data and normalized texture index for aluminum texture for 300 boxes of different shape (a-d) $M1$ and (e-h) $M2$

		E(sum)	TI(sum)	E(mean)	TI(mean)
$M1$ - iso	min(■)	0.25	1.06	0.25	1.06
	max(□)	2.04	5.12	1.92	4.63
$M2$ - iso	min(■)	0.24	1.05	0.24	1.05
	max(□)	2.08	5.30	1.94	4.73
$M1$ - alu	min(■)	0.24	1.07	0.24	0.89
	max(□)	2.00	4.57	1.67	4.12
$M2$ - alu	min(■)	0.25	1.05	0.25	0.85
	max(□)	1.90	4.28	1.69	4.20

Table 4.1: Extremal values associated with Figures 4.7 and 4.8

4.2 Clustering technique

Since measured orientation data often appears in the form of clouds of similar orientations in Euler space, another method to reduce data by using a cluster analysis is presented. Therefore, an algorithm being based on the k-means algorithm from Lloyd (1982) and its enhancement k-means++ (Arthur and Vassilvitskii, 2007) is applied. To use this algorithm, the existence of a distance measure between discrete data points is required. In the present case using orientation data, the distance between orientations is quantified using a definition of Bunge (1993) being based on the axis-angle-parametrization of an orientation, where the minimum angle ω is computed to transform an orientation Q_1 into another one Q_2 (see Eq. (2.45)).

The number of cluster centers is denoted by N, which in fact defines the degree of compression of the raw data and, therefore, is of significant importance for the resulting reduced data set. The initialization of the N clusters is performed in a way according to the k-means++-algorithm, which guarantees the result of the clustering process not to be governed by chance or by outlier values, as could be the case using the original k-means-procedure (Arthur and Vassilvitskii, 2007). The first cluster center is chosen randomly among the given orientation data points in Euler space. The next center is chosen so that a probability

$$P(Q) = \frac{\omega^2(Q)}{\sum_{\tilde{Q}} \omega^2(\tilde{Q})} \tag{4.15}$$

is maximized. In Eq. 4.15, $\omega(Q)$ denotes the distance to the nearest of the already

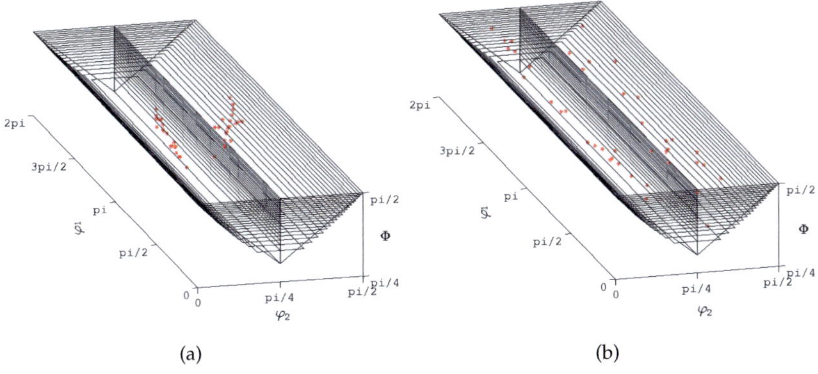

Figure 4.9: Texture component approximation by (a) sum and (b) mean for method $M1$ and the uniform texture distribution for $I - J - K = 1 - 20 - 2$

determined cluster centers, and the sum in the denominator is evaluated for all discrete orientations of the data set. After the initialization of all centers, each of the discrete orientations is assigned to the nearest cluster center. The mean orientation of the cloud of points is determined accounting for different weights of single orientations and the inhomogeneous metric of the orientation space using quaternions (as has already been described in the previous section). The mean orientation acts as the new cluster center. This procedure is repeated up to convergence (Fig. 4.10).

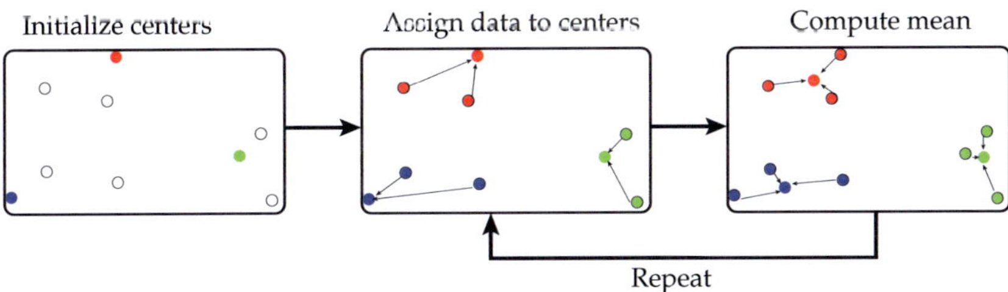

Figure 4.10: Schematic description of the k-means algorithm

The orientations of the converged cluster centers with their weights are the texture components of the reduced data set. Exemplarily, Fig 4.11 shows the classification of discrete data to clusters and Fig. 4.12 visualizes the iterative steps up to convergence of these cluster centers.

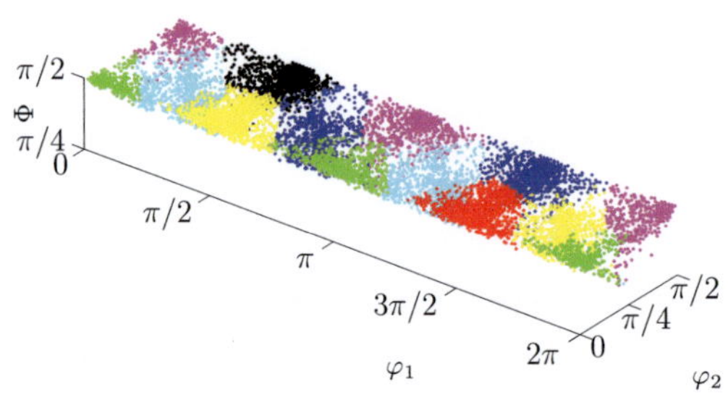

Figure 4.11: Discrete orientations of the raw aluminum texture data set assigned to 12 clusters

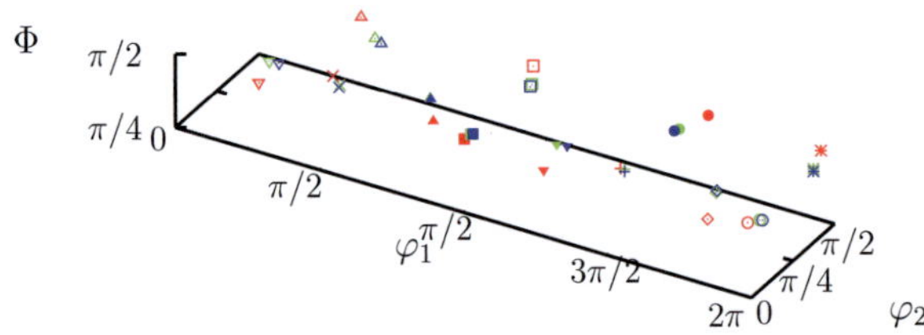

Figure 4.12: Example of iterative movement of 12 cluster centers up to convergence; colors denote: Initialized centers, first iteration, second iteration

4.2.1 Influence of number of clusters

Similar to the study in section 4.1.2, the prescribed number of clusters is varied to study its impact on the CODF approximation error and the texture index.

Once more, the two data sets 1) uniform orientation distribution and 2) cube-dominated aluminum texture are analyzed. Additionally, the two methods of averaging being called sum and mean are used. Figures 4.13 and 4.14 show the results of this study[6]. As expected due to the nature of the method, the

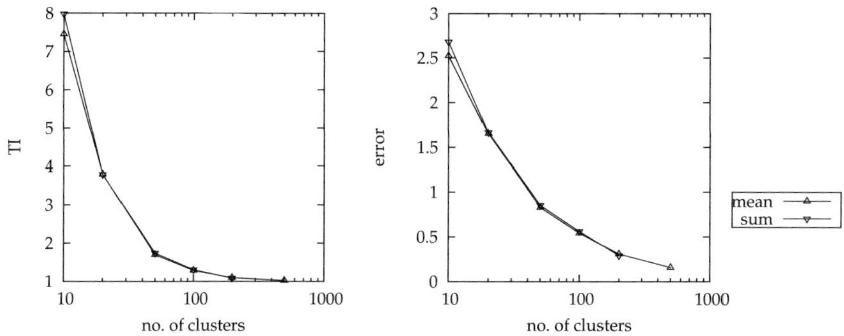

Figure 4.13: Normalized texture index (TI) and $L2$-error of the CODFs (error) for the uniform orientation distribution with respect to the number of clusters

clustering technique works better for textured orientation data compared to a uniform distribution. For the gray texture, the error in the CODFs is higher and does not decrease as rapidly as for the aluminum data set with increase of number of clusters. The results are nearly independent of the chosen averaging technique sum or mean. Nevertheless, small differences are observable, which can also be a resultant of different numbers of iteration steps needed to fulfill the termination criterion (distance of clusters in two subsequent iterations being smaller than a threshold value) of the algorithm in case of the two averaging schemes.

Since the initialization step in the case of kmeans++ takes most of the whole computational time of the clustering in the present study, the random initialization of the clusters according to the original kmeans procedure is tested (Fig. 4.15). In this case, a fixed number of three iteration steps are performed after initialization. The results obtained by random initialization (Fig. 4.15) are pretty similar

---

[6]Here, only results up to 500 clusters are shown, since this computation takes already 2.5 days and would even take longer for more cluster centers.

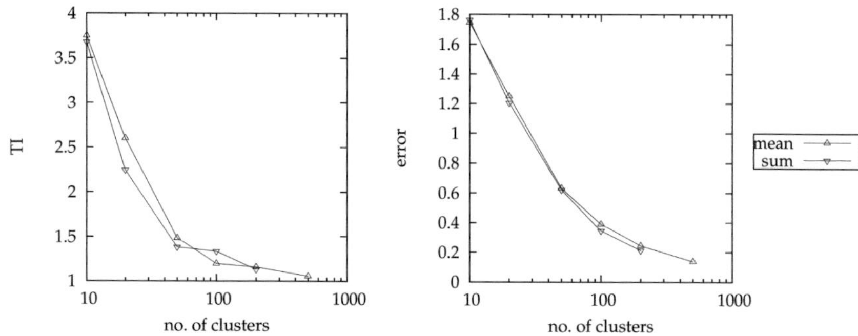

Figure 4.14: Normalized texture index (TI) and $L2$-error of the CODFs (error) for the aluminum texture with respect to the number of clusters

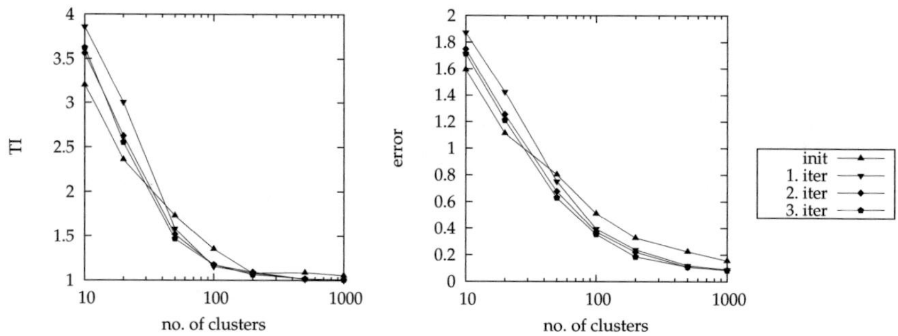

Figure 4.15: Results of clustering with random initialization of clusters and three iterations using mean averaging for the aluminum texture

to the results obtained after kmeans++ initialization (Fig. 4.14). However, they are generated in a fraction of time. It is, therefore, recommendable to use the random initialization instead of the advanced one, since if the results are not satisfying, the clustering could be redone with another random initialization and lots of time could be saved compared to the application of the kmeans++ procedure.

4.3 Comparison of the methods

The box method and the cluster method are both applicable for the representative reduction of orientation data sets. The box method is faster than the cluster method, since only the latter requires the computation of distances from each of the discrete orientations to all cluster centers in every iteration step. The cluster method is effective for a small number of clusters and yields better results than the box method for the same rate of compression, if the number of clusters is known a priori. Due to the data set independent choice of box boundaries, point clouds of orientation data could be artificially split into several boxes so that the compressed result is worse compared to the cluster method. However, this way of compression, being independent of the investigated data set, renders the algorithm work fast and it shows to give similar results as the cluster method when using a sufficient number of boxes. If it is a priori known that the measured data shows non-spherical clouds of data, in case of the clustering technique, the distance computation should be improved using an anisotropic metric. Otherwise, the box method gives better results than the cluster method for the same amount of compression. In general, the cluster mechanism generates convex clusters due to the distance criteria, which is not a problem for the usage with orientation data sets, if a sufficient number of clusters is used. In case of uniform orientation distribution, the box method inherently yields better results than the algorithm based on cluster detection. The cluster method is technically more simple, but due to the large amount of distance computations between individual data points, it takes a multiple of computational time compared to the box method. The box method is, therefore, recommended especially for reducing large data sets.

Chapter 5

Physically Linear Homogenization Methods

5.1 Localization tensors and effective elasticity tensors

In linear elasticity, the solution of an initial boundary value problem is unique. The stress and strain fields depend linearly on the loading, so that they can be represented as

$$
\begin{aligned}
\varepsilon(x) &= \mathbb{A}(x)[\bar{\varepsilon}] \quad \text{for} \quad u(x) = \bar{\varepsilon}x \quad \forall x \in \partial V, \\
\sigma(x) &= \mathbb{B}(x)[\bar{\sigma}] \quad \text{for} \quad t(x) = \bar{\sigma}n(x) \quad \forall x \in \partial V.
\end{aligned}
\tag{5.1}
$$

The tensors $\mathbb{A}$ and $\mathbb{B}$ are called localization tensors. They depend on the microstructure of the material and totally incorporate the solution of the boundary value problem.

Since the exact derivation of the localization tensors is, in general, impossible using analytical methods, a simple approximation is to work with phase averages of the fields

$$
\begin{aligned}
\varepsilon_\alpha &= \langle \varepsilon(x) \rangle_\alpha = \langle \mathbb{A}(x) \rangle_\alpha [\bar{\varepsilon}] = \mathbb{A}_\alpha[\bar{\varepsilon}], \\
\sigma_\alpha &= \langle \sigma(x) \rangle_\alpha = \langle \mathbb{B}(x) \rangle_\alpha [\bar{\sigma}] = \mathbb{B}_\alpha[\bar{\sigma}],
\end{aligned}
\tag{5.2}
$$

wherein $\mathbb{A}_\alpha$ and $\mathbb{B}_\alpha$ are the phase averages of the localization tensors. By Eqs. (5.1) and (5.2) it becomes obvious that localization tensors possess the minor symmetries ($\mathbb{A} = \mathbb{A}^{\mathsf{T_L}} = \mathbb{A}^{\mathsf{T_R}}$, $\mathbb{B} = \mathbb{B}^{\mathsf{T_L}} = \mathbb{B}^{\mathsf{T_R}}$) but in general not the major symmetry of fourth-order tensors ($\mathbb{A} \neq \mathbb{A}^{\mathsf{T}}$, $\mathbb{B} \neq \mathbb{B}^{\mathsf{T}}$). The volume average of localization tensors for a N-phase material with the volume fraction of phase α being c_α is given by

$$
\langle \mathbb{A}(x) \rangle = \sum_{\alpha=1}^{N} c_\alpha \mathbb{A}_\alpha = \mathbb{I}^S, \quad \langle \mathbb{B}(x) \rangle = \sum_{\alpha=1}^{N} c_\alpha \mathbb{B}_\alpha = \mathbb{I}^S.
\tag{5.3}
$$

The effective stiffness and compliance tensors, $\bar{\mathbb{C}}$ and $\bar{\mathbb{S}}$, respectively, relating the effective stress and strain fields of an N-phase material are given as

$$\begin{aligned}
\bar{\boldsymbol{\sigma}} &= \langle \boldsymbol{\sigma}(\boldsymbol{x}) \rangle = \langle \mathbb{C}(\boldsymbol{x})[\boldsymbol{\varepsilon}(\boldsymbol{x})] \rangle = \langle \mathbb{C}(\boldsymbol{x})\mathbb{A}(\boldsymbol{x}) \rangle [\bar{\boldsymbol{\varepsilon}}] = \bar{\mathbb{C}}[\bar{\boldsymbol{\varepsilon}}] &\rightarrow& \quad \bar{\mathbb{C}} = \langle \mathbb{C}\mathbb{A} \rangle, \\
\bar{\boldsymbol{\varepsilon}} &= \langle \boldsymbol{\varepsilon}(\boldsymbol{x}) \rangle = \langle \mathbb{S}(\boldsymbol{x})[\boldsymbol{\sigma}(\boldsymbol{x})] \rangle = \langle \mathbb{S}(\boldsymbol{x})\mathbb{B}(\boldsymbol{x}) \rangle [\bar{\boldsymbol{\sigma}}] = \bar{\mathbb{S}}[\bar{\boldsymbol{\sigma}}] &\rightarrow& \quad \bar{\mathbb{S}} = \langle \mathbb{S}\mathbb{B} \rangle.
\end{aligned} \tag{5.4}$$

On the other hand, starting from the energetic point of view that the elastic energy on the macro scale equals the volume average of the microscopic elastic energy

$$2\langle W \rangle = \langle \boldsymbol{\sigma}(\boldsymbol{x}) \cdot \boldsymbol{\varepsilon}(\boldsymbol{x}) \rangle = \bar{\boldsymbol{\sigma}} \cdot \bar{\boldsymbol{\varepsilon}} = 2\bar{W}, \tag{5.5}$$

the effective elasticity tensors are given by

$$\begin{aligned}
\langle \boldsymbol{\varepsilon}(\boldsymbol{x}) \cdot \mathbb{C}(\boldsymbol{x})[\boldsymbol{\varepsilon}(\boldsymbol{x})] \rangle &= \bar{\boldsymbol{\varepsilon}} \cdot \langle \mathbb{A}^{\mathsf{T}}(\boldsymbol{x})\mathbb{C}(\boldsymbol{x})\mathbb{A}(\boldsymbol{x}) \rangle [\bar{\boldsymbol{\varepsilon}}] = \bar{\boldsymbol{\varepsilon}} \cdot \bar{\mathbb{C}}[\bar{\boldsymbol{\varepsilon}}] &\rightarrow& \quad \bar{\mathbb{C}} = \langle \mathbb{A}^{\mathsf{T}}\mathbb{C}\mathbb{A} \rangle, \\
\langle \boldsymbol{\sigma}(\boldsymbol{x}) \cdot \mathbb{S}(\boldsymbol{x})[\boldsymbol{\sigma}(\boldsymbol{x})] \rangle &= \bar{\boldsymbol{\sigma}} \cdot \langle \mathbb{B}^{\mathsf{T}}(\boldsymbol{x})\mathbb{S}(\boldsymbol{x})\mathbb{B}(\boldsymbol{x}) \rangle [\bar{\boldsymbol{\sigma}}] = \bar{\boldsymbol{\sigma}} \cdot \bar{\mathbb{S}}[\bar{\boldsymbol{\sigma}}] &\rightarrow& \quad \bar{\mathbb{S}} = \langle \mathbb{B}^{\mathsf{T}}\mathbb{S}\mathbb{B} \rangle.
\end{aligned} \tag{5.6}$$

The two versions of defining the effective elasticity tensors (Eqs. (5.4) and (5.6)) are only equivalent, if the Hill condition (Eq. (2.57)) holds. In this case, the major symmetry and the positive definiteness of the effective stiffness tensors can be easily shown.[7]

5.2 Single inclusion problem

Since many linear homogenization schemes are based on Eshelbys pioneering work (Eshelby, 1957) solving the basic problem of one inhomogeneity/inclusion in an infinite matrix, this method is briefly reviewed in the following subsections.

5.2.1 Ellipsoidal inclusion in infinite matrix

An inclusion Ω in an infinitely expanded matrix, both possessing the same elastic material behavior with stiffness $\mathbb{C}$, is assumed to be submitted to a uniform eigenstrain $\boldsymbol{\varepsilon}^* \neq 0$ (Fig. 5.1(a)). The eigenstrains can be, e.g., plastic strains, strains due to thermal expansion, initial strains, etc. The strain field is,

---

[7]Note the work of Benveniste et al. (1991), who discuss the loss of major symmetry of some approximations for effective stiffness tensors by mean-field approaches.

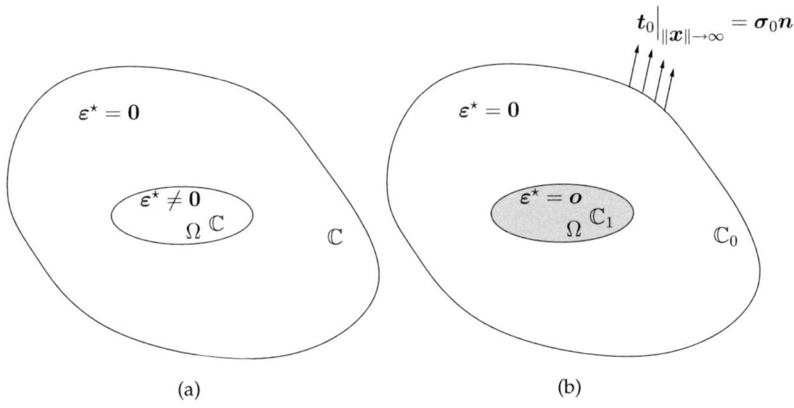

Figure 5.1: (a) Single inclusion Ω in infinite matrix, (b) single inhomogeneity Ω in infinite matrix subjected to Neumann boundary conditions

therefore, composed of elastic and eigenstrains

$$\varepsilon(\boldsymbol{x}) = \varepsilon^e(\boldsymbol{x}) + \varepsilon^\star(\boldsymbol{x}), \quad \varepsilon(\boldsymbol{x}) = \mathrm{sym}(\boldsymbol{H}(\boldsymbol{x})), \quad \boldsymbol{H}(\boldsymbol{x}) = \mathrm{grad}\,(\boldsymbol{u}(\boldsymbol{x})), \qquad (5.7)$$

so that Hooke's law reads

$$\boldsymbol{\sigma}(\boldsymbol{x}) = \mathbb{C}[\varepsilon(\boldsymbol{x}) - \varepsilon^\star(\boldsymbol{x})] = \mathbb{C}[\boldsymbol{H}(\boldsymbol{x}) - \varepsilon^\star(\boldsymbol{x})]. \qquad (5.8)$$

The static equilibrium equation is then given by

$$\mathrm{div}\,(\boldsymbol{\sigma}(\boldsymbol{x})) = \mathrm{div}\,(\mathbb{C}[\boldsymbol{H}](\boldsymbol{x})) + \mathrm{div}\,(-\mathbb{C}[\varepsilon^\star(\boldsymbol{x})]) = \boldsymbol{o}, \qquad (5.9)$$

where $\boldsymbol{b} = \mathrm{div}\,(-\,\mathbb{C}[\varepsilon^\star])$ can be interpreted as a body force. In order to solve for the displacement field yielding the equilibrated state, Fourier transformation is used for transferring the partial differential equations into an algebraic equation. For Fourier transforming a vector, e.g., the displacement $\boldsymbol{u}$, giving its Fourier transform $\hat{\boldsymbol{u}}$, the following rules hold[8]

$$\hat{\boldsymbol{u}}(\boldsymbol{\xi}) = \frac{1}{(2\pi)^3} \iiint\limits_{-\infty}^{\infty} \boldsymbol{u}(\boldsymbol{x}) \exp(-i\boldsymbol{x}\cdot\boldsymbol{\xi})\,\mathrm{d}\boldsymbol{x}, \qquad (5.10)$$

$$\boldsymbol{u}(\boldsymbol{x}) = \iiint\limits_{-\infty}^{\infty} \hat{\boldsymbol{u}}(\boldsymbol{\xi}) \exp(i\boldsymbol{x}\cdot\boldsymbol{\xi})\,\mathrm{d}\boldsymbol{\xi}. \qquad (5.11)$$

---

[8]Note that there exist different definitions of Fourier transformations. Sometimes, a factor $1/(2\pi)^{3/2}$ is used in both, the Fourier transform and the inverse operation or on the other hand, the factor $1/(2\pi)^3$ could also only appear in the inverse transform.

Analogously, the components of the Fourier transform of $\boldsymbol{H} = \mathrm{grad}\,(\boldsymbol{u}(\boldsymbol{x})) = \partial u_i(\boldsymbol{x})/\partial x_j\, \boldsymbol{e}_i \otimes \boldsymbol{e}_j$ are given by

$$\hat{H}_{ij}(\boldsymbol{\xi}) = \frac{1}{(2\pi)^3}\iiint\limits_{-\infty}^{\infty}\frac{\partial u_i(\boldsymbol{x})}{\partial x_j}\exp(-ix_l\xi_l)\,\mathrm{d}\boldsymbol{x} = i\xi_j\hat{u}_i(\boldsymbol{\xi}), \tag{5.12}$$

$$\hat{\boldsymbol{H}}(\boldsymbol{\xi}) = i\hat{\boldsymbol{u}}(\boldsymbol{\xi})\otimes\boldsymbol{\xi}, \tag{5.13}$$

at which it has already been taken advantage of the fact that $\boldsymbol{u}(\boldsymbol{x})\big|_{\|\boldsymbol{x}\|\to\infty} = \boldsymbol{o}$ holds due to only a finite region being subjected to eigenstrains. For the same reason, the Fourier transform of the components of $\mathrm{grad}\,(\mathrm{grad}\,(\boldsymbol{u}(\boldsymbol{x})))$ is given by

$$\frac{1}{(2\pi)^3}\iiint\limits_{-\infty}^{\infty}\frac{\partial^2 u_i(\boldsymbol{x})}{\partial x_j\partial x_k}\exp(-ix_l\xi_l)\,\mathrm{d}\boldsymbol{x} = -\xi_j\xi_k\hat{u}_i(\boldsymbol{\xi}), \tag{5.14}$$

and in tensorial form

$$\frac{1}{(2\pi)^3}\iiint\limits_{-\infty}^{\infty}\mathrm{grad}\,(\mathrm{grad}\,(\boldsymbol{u}(\boldsymbol{x})))\exp(-i\boldsymbol{x}\cdot\boldsymbol{\xi})\,\mathrm{d}\boldsymbol{x} = -\hat{\boldsymbol{u}}(\boldsymbol{\xi})\otimes\boldsymbol{\xi}\otimes\boldsymbol{\xi}. \tag{5.15}$$

Using Eqs. (5.10)- (5.15), the Fourier transform of the equilibrium equation (5.9) reads

$$\boldsymbol{K}(\boldsymbol{\xi})\hat{\boldsymbol{u}}(\boldsymbol{\xi}) = i\mathbb{C}[\boldsymbol{\xi}\otimes\hat{\boldsymbol{\varepsilon}}^{\star}(\boldsymbol{\xi})] = \hat{\boldsymbol{b}}, \quad \boldsymbol{K} = \mathbb{C}[[\boldsymbol{\xi}\otimes\boldsymbol{\xi}]], \tag{5.16}$$

so that

$$\boldsymbol{u}(\boldsymbol{x}) = \iiint\limits_{-\infty}^{\infty}\boldsymbol{K}^{-1}(\boldsymbol{\xi})\hat{\boldsymbol{b}}(\boldsymbol{\xi})\exp(i\boldsymbol{\xi}\cdot\boldsymbol{x})\,\mathrm{d}\boldsymbol{\xi}. \tag{5.17}$$

With the inverse Fourier transform of the eigenstrains, the displacement field is

$$\boldsymbol{u}(\boldsymbol{x}) = -\frac{1}{(2\pi)^3}\mathrm{div}_{\boldsymbol{x}}\left(\iiint\limits_{-\infty}^{\infty}\iiint\limits_{-\infty}^{\infty}\boldsymbol{K}^{-1}(\boldsymbol{\xi})(\mathbb{C}[\boldsymbol{\varepsilon}^{\star}(\boldsymbol{y})])\exp(i\boldsymbol{\xi}\cdot(\boldsymbol{x}-\boldsymbol{y}))\,\mathrm{d}\boldsymbol{\xi}\,\mathrm{d}\boldsymbol{y}\right) \tag{5.18}$$

or, equivalently,

$$u_i(\boldsymbol{x}) = -\frac{1}{(2\pi)^3}\frac{\partial}{\partial x_k}\iiint\limits_{-\infty}^{\infty}\iiint\limits_{-\infty}^{\infty}K_{ij}^{-1}(\boldsymbol{\xi})C_{jklm}\varepsilon_{lm}^{\star}(\boldsymbol{y})\exp(i\xi_n(x_n-y_n))\,\mathrm{d}\boldsymbol{\xi}\,\mathrm{d}\boldsymbol{y}. \tag{5.19}$$

At this point, usually the infinite body's Green's function is introduced

$$\boldsymbol{G}(\boldsymbol{x}-\boldsymbol{y}) = \frac{1}{(2\pi)^3}\iiint\limits_{-\infty}^{\infty}\boldsymbol{K}^{-1}(\boldsymbol{\xi})\exp(i\boldsymbol{\xi}\cdot(\boldsymbol{x}-\boldsymbol{y}))\,\mathrm{d}\boldsymbol{\xi}, \tag{5.20}$$

so that

$$u(x) = - \iiint_{-\infty}^{\infty} \operatorname{grad}_x \left(G(x - y) \right) \left[\mathbb{C}[\varepsilon^\star(y)] \right] \mathrm{d}y \qquad (5.21)$$

or, equivalently,

$$u_i(x) = - \iiint_{-\infty}^{\infty} \frac{\partial G_{ij}(x - y)}{\partial x_k} C_{jklm}\varepsilon^\star_{lm}(y) \, \mathrm{d}y. \qquad (5.22)$$

Knowing that the integrand vanishes due to $\varepsilon^\star = o \; \forall x \notin \Omega$,

$$u(x) = - \iiint_{\Omega} \operatorname{grad}_x \left(G(x - y) \right) \left[\mathbb{C}[\varepsilon^\star(y)] \right] \mathrm{d}y \qquad (5.23)$$

holds, or with $\varepsilon^\star = \mathrm{const}$

$$u(x) = - \left(\iiint_{\Omega} \operatorname{grad}_x \left(G(x - y) \right) \, \mathrm{d}y \right) [\mathbb{C}[\varepsilon^\star]]. \qquad (5.24)$$

The strain in the inclusion then reads

$$\varepsilon(x) = -\operatorname{sym} \left(\operatorname{grad}_x \left(\left(\iiint_{\Omega} \left(\operatorname{grad}_x \left(G(x - y) \right) \right) \, \mathrm{d}y \right) [\mathbb{C}[\varepsilon^\star]] \right) \right). \qquad (5.25)$$

Introducing

$$\boldsymbol{\Gamma}(x - y) = -\frac{1}{4} \left(\frac{\partial^2 G_{ij}}{\partial x_k \partial x_n} + \frac{\partial^2 G_{ik}}{\partial x_j \partial x_n} + \frac{\partial^2 G_{nj}}{\partial x_k \partial x_i} + \frac{\partial^2 G_{nk}}{\partial x_j \partial x_i} \right) e_i \otimes e_n \otimes e_j \otimes e_k \qquad (5.26)$$

with $G = G(x - y)$, the solution for the strain can be given in a short form

$$\varepsilon(x) = \left(\iiint_{\Omega} \boldsymbol{\Gamma}(x - y) \, \mathrm{d}y \right) \mathbb{C}[\varepsilon^\star] = \mathbb{P}(x)\mathbb{C}[\varepsilon^\star] = \mathbb{E}(x)[\varepsilon^\star], \qquad (5.27)$$

with Hill's polarization tensor $\mathbb{P}$ and Eshelbys tensor $\mathbb{E}$. While Hill's polarization tensor describes the mapping of the stress due to eigenstrain on the strain field, Eshelbys tensor gives the influence of eigenstrain onto the strain field. Eshelby showed that if the domain Ω is of ellipsoidal shape, $\mathbb{P} = \mathbb{P}_0 = \mathrm{const}$, so that the strain field in the inclusion is constant. The proof for the uniformity of the polarization tensor in this case is given in section 5.2.3.

5.2.2 Ellipsoidal inhomogeneity in infinite matrix

A generalization of the single inclusion problem of the previous section is the investigation of an ellipsoidal inhomogeneity Ω with stiffness $\mathbb{C}_1$ in a matrix with stiffness $\mathbb{C}_0$ depicted in Fig. 5.1(b) (Eshelby, 1957). The compound is loaded by a stress σ_0 at infinity. The total stress can, therefore, be given as

$$\sigma(x) = \sigma_0 + \tilde{\sigma}(x), \tag{5.28}$$

where $\tilde{\sigma}$ denotes the stress fluctuation due to the presence of the inhomogeneity. Since the boundary stress is constant, the equilibrium equation reads $\operatorname{div}(\tilde{\sigma}) = o$ and, furthermore, $\tilde{\sigma}\big|_{\|x\|\to\infty} = 0$ must hold. Inside Ω, Hooke's law is given in the form

$$\sigma_0 + \tilde{\sigma}(x) = \mathbb{C}_1[\varepsilon_0 + \tilde{\varepsilon}(x)] \quad \forall x \in \Omega, \tag{5.29}$$

with $\varepsilon_0 = \operatorname{sym}(\operatorname{grad}(u_0))$, the strain associated with the displacement field u_0 being compatible to the boundary condition for a homogeneous body and the strain fluctuation $\tilde{\varepsilon}$ due to the inhomogeneity. To use the results from the previous section, a transition to a homogeneous body with an ellipsoidal domain with eigenstrain $\varepsilon^\star \neq 0$ has to be created (Fig. 5.1(b) to Fig. 5.1(a)), for which $\varepsilon^\star$ has to be determined appropriately. In this case, Eq. (5.29) becomes

$$\sigma_0 + \tilde{\sigma}(x) = \mathbb{C}_0[\varepsilon_0 + \tilde{\varepsilon}(x) - \varepsilon^\star(x)]. \tag{5.30}$$

Both problems are equivalent only if

$$\mathbb{C}_1[\varepsilon_0 + \tilde{\varepsilon}(x)] = \mathbb{C}_0[\varepsilon_0 + \tilde{\varepsilon}(x) - \varepsilon^\star(x)] \tag{5.31}$$

holds. If $\sigma_0 = \text{const}$, $\varepsilon^\star = \text{const}$ with $\tilde{\varepsilon} = \mathbb{E}[\varepsilon^\star]$, so that the eigenstrain due to the constraint Eq. (5.31) is determined as

$$\varepsilon^\star = -\left(\mathbb{E} + (\mathbb{C}_1 - \mathbb{C}_0)^{-1}\mathbb{C}_0\right)^{-1}[\varepsilon_0] \tag{5.32}$$

and the total strain inside the inhomogeneity is

$$\begin{aligned}
\varepsilon &= \left(\mathbb{E}\mathbb{C}_0^{-1}(\mathbb{C}_1 - \mathbb{C}_0) + \mathbb{I}\right)^{-1}[\varepsilon_0] \\
&= (\mathbb{P}_0(\mathbb{C}_1 - \mathbb{C}_0) + \mathbb{I})^{-1}[\varepsilon_0], \quad \forall x \in \Omega.
\end{aligned} \tag{5.33}$$

As Eq. (5.33) shows, the strain localization tensor mapping the macroscopic strain ($\bar{\varepsilon} = \varepsilon_0$) onto the phase average strain of the inhomogeneity $\varepsilon = \mathbb{A}[\bar{\varepsilon}]$ is given by

$$\mathbb{A} = \left(\mathbb{I} + \mathbb{E}\mathbb{C}_0^{-1}(\mathbb{C}_1 - \mathbb{C}_0)\right)^{-1} = (\mathbb{I} + \mathbb{P}_0(\mathbb{C}_1 - \mathbb{C}_0))^{-1}. \tag{5.34}$$

While $\mathbb{E}$, $\mathbb{A}$ and $\mathbb{P}_0$ possess the left and right minor symmetries, in general only $\mathbb{P}_0$ additionally possesses the major symmetry.

5.2.3 Hills polarization tensor for ellipsoidal domain

In this section, Hills polarization tensor is derived on the basis of the works of Mura (1987) and Qu and Cherkaoui (2006) for the general case of an anisotropic ellipsoidal inclusion or inhomogeneity, respectively.

The ellipsoid Ω (see Fig. 5.2) with the half-axes a_1, a_2 and a_3 being aligned with the coordinate system, is described by

$$\Omega = \{\boldsymbol{x}; \boldsymbol{x} \cdot (\boldsymbol{Z}\boldsymbol{x}) \leq 1\} \quad \text{with} \quad \boldsymbol{Z} = \sum_{i=1}^{3} a_i^{-2}\boldsymbol{e}_i \otimes \boldsymbol{e}_i. \tag{5.35}$$

Infinite body's Green's function (Eq. (5.20)) can be computed using either $\boldsymbol{\xi}$ or

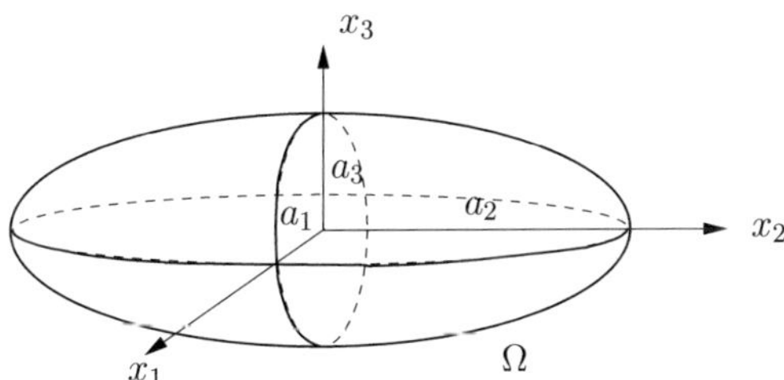

Figure 5.2: Ellipsoid

$-\boldsymbol{\xi}$, so that with $\boldsymbol{K}^{-1}$ being homogeneous of degree -2 in $\boldsymbol{\xi}$, Eq. (5.20) can be given in two equivalent forms

$$\begin{aligned} \boldsymbol{G}(\boldsymbol{x} - \boldsymbol{y}) &= \frac{1}{(2\pi)^3} \iiint\limits_{-\infty}^{\infty} \boldsymbol{K}^{-1}(\boldsymbol{\xi}) \exp(i\boldsymbol{\xi} \cdot (\boldsymbol{x} - \boldsymbol{y}))\, \mathrm{d}\boldsymbol{\xi} \\ &= \frac{1}{(2\pi)^3} \iiint\limits_{-\infty}^{\infty} \boldsymbol{K}^{-1}(\boldsymbol{\xi}) \exp(-i\boldsymbol{\xi} \cdot (\boldsymbol{x} - \boldsymbol{y}))\, \mathrm{d}\boldsymbol{\xi}. \end{aligned} \tag{5.36}$$

The differential volume element $\mathrm{d}\boldsymbol{\xi}$ in Eq. (5.36) can be replaced by $\mathrm{d}\boldsymbol{\xi} = \mathrm{d}\xi_1\, \mathrm{d}\xi_2\, \mathrm{d}\xi_3 = \mathrm{d}\xi\, \mathrm{d}S_\xi(\boldsymbol{\xi}) = \xi^2\, \mathrm{d}\xi\, \mathrm{d}S(\boldsymbol{n}_\xi)$, where $\xi = \|\boldsymbol{\xi}\|$ and $\boldsymbol{n}_\xi = \boldsymbol{\xi}/\xi$ hold.

Furthermore, dS_ξ is a surface element of a sphere with radius ξ and dS a surface element of the unit sphere S in the origin, respectively. Once more accounting for $\boldsymbol{K}^{-1}(\boldsymbol{\xi})$ being homogeneous of degree -2 in $\boldsymbol{\xi}$, Eq. (5.36) can be rewritten

$$
\begin{aligned}
\boldsymbol{G}(\boldsymbol{x}-\boldsymbol{y}) &= \frac{1}{(2\pi)^3} \int_0^\infty \iint_{\partial S} \boldsymbol{K}^{-1}(\xi\boldsymbol{n}_\xi)\exp(i\xi\boldsymbol{n}_\xi\cdot(\boldsymbol{x}-\boldsymbol{y}))\xi^2\,\mathrm{d}S(\boldsymbol{n}_\xi)\,\mathrm{d}\xi \\
&= \frac{1}{(2\pi)^3} \int_0^\infty \iint_{\partial S} \boldsymbol{K}^{-1}(\boldsymbol{n}_\xi)\exp(i\xi\boldsymbol{n}_\xi\cdot(\boldsymbol{x}-\boldsymbol{y}))\,\mathrm{d}S(\boldsymbol{n}_\xi)\,\mathrm{d}\xi.
\end{aligned} \tag{5.37}
$$

The same can be considered for $\xi \to -\xi$, so that $\xi \in (-\infty, 0]$, leading to

$$
\boldsymbol{G}(\boldsymbol{x}-\boldsymbol{y}) = \frac{1}{(2\pi)^3} \int_{-\infty}^0 \iint_{\partial S} \boldsymbol{K}^{-1}(\boldsymbol{n}_\xi)\exp(i\xi\boldsymbol{n}_\xi\cdot(\boldsymbol{x}-\boldsymbol{y}))\,\mathrm{d}S(\boldsymbol{n}_\xi)\,\mathrm{d}\xi. \tag{5.38}
$$

The sum of the two latter equations result in

$$
\begin{aligned}
2\boldsymbol{G}(\boldsymbol{x}-\boldsymbol{y}) &= \frac{1}{(2\pi)^3} \int_{-\infty}^\infty \iint_{\partial S} \boldsymbol{K}^{-1}(\boldsymbol{n}_\xi)\exp(i\xi\boldsymbol{n}_\xi\cdot(\boldsymbol{x}-\boldsymbol{y}))\,\mathrm{d}S(\boldsymbol{n}_\xi)\,\mathrm{d}\xi \\
&= \frac{1}{(2\pi)^3} \iint_{\partial S} \left(\int_{-\infty}^\infty \exp(i\xi\boldsymbol{n}_\xi\cdot(\boldsymbol{x}-\boldsymbol{y}))\,\mathrm{d}\xi \right) \boldsymbol{K}^{-1}(\boldsymbol{n}_\xi)\,\mathrm{d}S(\boldsymbol{n}_\xi).
\end{aligned} \tag{5.39}
$$

Using the definitions of the Fourier transform of the Dirac distribution and its inverse, respectively,

$$
\hat{\delta}(\xi) = \frac{1}{2\pi} \int_{-\infty}^\infty \delta(x)\exp(-i\xi x)\,\mathrm{d}x = \frac{1}{2\pi}, \tag{5.40}
$$

$$
\delta(x) = \int_{-\infty}^\infty \hat{\delta}(\xi)\exp(i\xi x)\,\mathrm{d}\xi = \frac{1}{2\pi} \int_{-\infty}^\infty \exp(i\xi x)\,\mathrm{d}\xi, \tag{5.41}
$$

the following identity

$$
2\pi\delta(\boldsymbol{n}_\xi\cdot(\boldsymbol{x}-\boldsymbol{y})) = \int_{-\infty}^\infty \exp(i\xi\boldsymbol{n}_\xi\cdot(\boldsymbol{x}-\boldsymbol{y}))\,\mathrm{d}\xi \tag{5.42}
$$

leads from Eq. (5.39) to a form for Green's function which only involves the integration over the unit sphere

$$
\boldsymbol{G}(\boldsymbol{x}-\boldsymbol{y}) = \frac{1}{8\pi^2} \iint_{\partial S} \delta(\boldsymbol{n}_\xi\cdot(\boldsymbol{x}-\boldsymbol{y}))\boldsymbol{K}^{-1}(\boldsymbol{n}_\xi)\,\mathrm{d}S(\boldsymbol{n}_\xi). \tag{5.43}
$$

In each of the summands appearing in Eq. (5.27), being the integral over the ellipsoid of second derivatives of Green's functions, the order of integration can be exchanged, so that, e.g.,

$$\frac{\partial^2}{\partial x_k \partial x_n} \iiint_\Omega G_{ij}(\boldsymbol{x} - \boldsymbol{y})\, \mathrm{d}\boldsymbol{y} =$$

$$\frac{1}{8\pi^2} \frac{\partial^2}{\partial x_k \partial x_n} \iint_{\partial \mathcal{S}} \left(\iiint_\Omega \delta(\boldsymbol{n}_\xi \cdot (\boldsymbol{x} - \boldsymbol{y}))\, \mathrm{d}\boldsymbol{y} \right) K_{ij}^{-1}(\boldsymbol{n}_\xi)\, \mathrm{d}S(\boldsymbol{n}_\xi). \qquad (5.44)$$

For converting the integral over the ellipsoid Ω in Eq. (5.44) into an integral over the unit sphere $\mathcal{S}$, the following transformations are used[9]

$$\bar{\boldsymbol{y}} = \sqrt{\boldsymbol{Z}}\boldsymbol{y}, \quad \mathrm{d}\boldsymbol{y} = \det(\sqrt{\boldsymbol{Z}^{-1}})\, \mathrm{d}\bar{\boldsymbol{y}},$$
$$\bar{\boldsymbol{x}} = \sqrt{\boldsymbol{Z}}\boldsymbol{x}, \quad \bar{\boldsymbol{n}}_\xi = \sqrt{\boldsymbol{Z}^{-1}}\boldsymbol{n}_\xi / \|\sqrt{\boldsymbol{Z}^{-1}}\boldsymbol{n}_\xi\|. \qquad (5.45)$$

Thus,

$$\iiint_\Omega \delta(\boldsymbol{n}_\xi \cdot (\boldsymbol{x} - \boldsymbol{y}))\, \mathrm{d}\boldsymbol{y} = \iiint_\mathcal{S} \delta(\|\sqrt{\boldsymbol{Z}^{-1}}\boldsymbol{n}_\xi\|\bar{\boldsymbol{n}}_\xi \cdot (\bar{\boldsymbol{x}} - \bar{\boldsymbol{y}}))\det(\sqrt{\boldsymbol{Z}^{-1}})\, \mathrm{d}\bar{\boldsymbol{y}} \qquad (5.46)$$

can be integrated for inner points of the ellipsoid $|\bar{\boldsymbol{n}}_\xi \cdot \bar{\boldsymbol{x}}| < 1$ using cylindrical coordinates $\{r, \varphi, z\}$, with the z-axis being aligned with $\bar{\boldsymbol{n}}_\xi$. For an improved readability, the abbreviations $\alpha = \|\sqrt{\boldsymbol{Z}^{-1}}\boldsymbol{n}_\xi\|$ and $\beta = \det(\sqrt{\boldsymbol{Z}^{-1}})$ are introduced, so that

$$\iiint_\mathcal{S} \delta(\alpha\bar{\boldsymbol{n}}_\xi \cdot (\bar{\boldsymbol{x}} - \bar{\boldsymbol{y}}))\beta\, \mathrm{d}\bar{\boldsymbol{y}} - \beta \int_0^{2\pi} \int_{-1}^1 \int_0^{\sqrt{1-z^2}} \delta(\alpha(\bar{\boldsymbol{n}}_\xi \cdot \bar{\boldsymbol{x}} - z))r\, \mathrm{d}r\, \mathrm{d}z\, \mathrm{d}\varphi$$

$$= \frac{\beta}{\alpha} \int_0^{2\pi} \int_{-1}^1 \int_0^{\sqrt{1-z^2}} \delta(\bar{\boldsymbol{n}}_\xi \cdot \bar{\boldsymbol{x}} - z)r\, \mathrm{d}r\, \mathrm{d}z\, \mathrm{d}\varphi$$

$$= \frac{\pi\beta}{\alpha} \int_{-1}^1 \delta(\bar{\boldsymbol{n}}_\xi \cdot \bar{\boldsymbol{x}} - z))(1 - z^2)\, \mathrm{d}z$$

$$= \frac{\pi\beta}{\alpha}(1 - (\bar{\boldsymbol{n}}_\xi \cdot \bar{\boldsymbol{x}})^2)$$

$$= \frac{\pi\beta}{\alpha^3}(\alpha^2 - (\boldsymbol{n}_\xi \cdot \boldsymbol{x})^2). \qquad (5.47)$$

---

[9]Note that if Ω is not aligned with the axis of the coordinate system, then $\boldsymbol{Z}$ has to be spectrally decomposed in order to compute $\sqrt{\boldsymbol{Z}}$.

The substitution of Eq. (5.47) into Eq. (5.44) and further on into Eq. (5.27) yields Hill's polarization tensor

$$\mathbb{P}(\boldsymbol{x}) = \iiint_{\Omega} \boldsymbol{\Gamma}(\boldsymbol{x} - \boldsymbol{y}) \, \mathrm{d}\boldsymbol{y} \tag{5.48}$$

$$= -\frac{\beta}{8\pi} \iint_{\partial S} \alpha^{-3} \mathbb{I}^S \left(\frac{\partial^2 (\alpha^2 - (\boldsymbol{n}_\xi \cdot \boldsymbol{x})^2)}{\partial \boldsymbol{x}^2} \Box \boldsymbol{K}^{-1}(\boldsymbol{n}_\xi) \right) \mathbb{I}^S \, \mathrm{d}S(\boldsymbol{n}_\xi) \tag{5.49}$$

$$= \frac{\beta}{4\pi} \iint_{\partial S} \alpha^{-3} \mathbb{I}^S \left((\boldsymbol{n}_\xi \otimes \boldsymbol{n}_\xi) \Box \boldsymbol{K}^{-1}(\boldsymbol{n}_\xi) \right) \mathbb{I}^S \, \mathrm{d}S(\boldsymbol{n}_\xi) := \mathbb{P}_0 = \mathrm{const.} \tag{5.50}$$

By reversing the substitutions α and β, as well as using $\boldsymbol{A} = \sqrt{\boldsymbol{Z}}$, Hill's polarization tensor in its well-known form is obtained (see, e.g., Willis, 1977, 1981; Ponte Castañeda and Suquet, 1998; Jöchen and Böhlke, 2012)

$$\mathbb{P}_0 = \frac{1}{4\pi \det(\boldsymbol{A})} \iint_{\partial S} \mathbb{H}(\mathbb{C}, \boldsymbol{n})(\boldsymbol{n} \cdot (\boldsymbol{A}^{-\mathsf{T}} \boldsymbol{A}^{-1} \boldsymbol{n}))^{-3/2} \, \mathrm{d}S(\boldsymbol{n}),$$

$$\mathbb{H} = \mathbb{I}^S \left((\boldsymbol{n} \otimes \boldsymbol{n}) \Box \boldsymbol{K}^{-1}(\boldsymbol{n}) \right) \mathbb{I}^S, \quad \boldsymbol{K} = \mathbb{C}[[\boldsymbol{n} \otimes \boldsymbol{n}]]. \tag{5.51}$$

In general, $\mathbb{P}_0$ needs to be computed numerically. For example for an equiaxed morphology of the grains ($\boldsymbol{A} = \boldsymbol{I}$), Hills polarization tensor simplifies to

$$\mathbb{P}_0(\mathbb{C}) = \frac{1}{4\pi} \iint_{\partial S} \mathbb{H}(\mathbb{C}, \boldsymbol{n}) \, \mathrm{d}S(\boldsymbol{n}), \tag{5.52}$$

nevertheless generally not being integrable in closed form. For some special cases, analytical solutions exist, e.g., for an isotropic material stiffness $\mathbb{C}$ and a spheroidal shape (oblate and prolate) of the inclusion (see Willis (1981)) or a spherical inclusion in an anisotropic matrix (see Ponte Castañeda and Suquet (1998)). For the most simple case of spherical inclusion in an isotropic material $\mathbb{C} = (3\lambda + 2\mu)\mathbb{P}_1^I + 2\mu\mathbb{P}_2^I$ with Lamé constants λ and μ, where also the bulk modulus $3K = 3\lambda + 2\mu$ can be used, the acoustic tensor $\boldsymbol{K}$ and the inverse $\boldsymbol{K}^{-1}$ are given by

$$\boldsymbol{K}(\boldsymbol{n}) = (\lambda + \mu)\boldsymbol{n} \otimes \boldsymbol{n} + \mu\boldsymbol{I},$$

$$\boldsymbol{K}^{-1}(\boldsymbol{n}) = -\frac{\lambda + \mu}{\mu(\lambda + 2\mu)} \boldsymbol{n} \otimes \boldsymbol{n} + \frac{1}{\mu}\boldsymbol{I}, \tag{5.53}$$

so that

$$\mathbb{H} = -\frac{\lambda + \mu}{\mu(\lambda + 2\mu)} \boldsymbol{n} \otimes \boldsymbol{n} \otimes \boldsymbol{n} \otimes \boldsymbol{n} + \frac{1}{\mu} \mathbb{I}^S (\boldsymbol{I} \Box (\boldsymbol{n} \otimes \boldsymbol{n}) \mathbb{I}^S. \tag{5.54}$$

Integrating the fourth-order tensors in Eq. (5.54) over the unit sphere

$$\iint_{\partial \mathcal{S}} \boldsymbol{n} \otimes \boldsymbol{n} \otimes \boldsymbol{n} \otimes \boldsymbol{n} \, \mathrm{d}S(\boldsymbol{n}) = \frac{4\pi}{15}\left(3\mathbb{P}_1^I + 2\mathbb{I}^S\right), \tag{5.55}$$

$$\iint_{\partial \mathcal{S}} \mathbb{I}^S \left(\boldsymbol{I} \square (\boldsymbol{n} \otimes \boldsymbol{n})\right) \mathbb{I}^S \, \mathrm{d}S(\boldsymbol{n}) = \frac{4\pi}{3}\mathbb{I}^S, \tag{5.56}$$

Hill's polarization tensor for that special case of spherical inclusion in isotropic material is given by

$$\mathbb{P}_0 = \frac{1}{3(\lambda + 2\mu)}\mathbb{P}_1^I + \frac{3\lambda + 8\mu}{15\mu(\lambda + 2\mu)}\mathbb{P}_2^I = \frac{1}{3K + 4\mu}\mathbb{P}_1^I + \frac{3(K + 2\mu)}{5\mu(3K + 4\mu)}\mathbb{P}_2^I. \tag{5.57}$$

5.3 Bounds

Characterizing precisely the effective behavior of a heterogeneous material is usually difficult. Approximations (see, e.g., sections 5.4 and 5.5) can be obtained, the results of which strongly depend on their assumptions, so that it is nearly impossible to judge the quality of the approximation. Bounding the effective material is, therefore, a useful approach to identify the interval of the potential material response. It depends on the amount and type of anisotropy to decide whether it is sufficient to evaluate the elementary bounds or if higher-order bounds are required.

5.3.1 First-order bounds

Voigt (1889) suggested to assume homogeneous strain $\varepsilon(x) = \bar{\varepsilon}$ and Reuss (1929) homogeneous stress $\sigma(x) = \bar{\sigma}$ throughout the heterogeneous material, which can be shown to give upper bounds for the elastic strain energy and complementary energy, respectively (Hill, 1952). Obviously, the Voigt assumption fulfills kinematic compatibility while accepting the loss of statical compatibility. The Reuss method prioritizes exactly the other way round. In real microstructures, both compatibilities are fulfilled, so that the extremal cases of Voigt and Reuss are never reached in reality but the behavior definitely lies in between these bounds.

Due to the aforementioned assumptions, the strain localization tensor in the Voigt case becomes $\mathbb{A}^V = \mathbb{I}^S$, while the stress localization tensor in the Reuss method is trivially $\mathbb{B}^R = \mathbb{I}^S$, so that the effective stiffnesses for these methods are simply the arithmetic and harmonic mean of the stiffnesses of the single phases and read

$$\bar{\mathbb{C}}^V = \langle \mathbb{C}(\boldsymbol{x}) \rangle = \frac{1}{V} \int_V \mathbb{C}(\boldsymbol{x}) \, \mathrm{d}V = \int_Q f(\boldsymbol{Q}) \mathbb{C}(\boldsymbol{Q}) \, \mathrm{d}Q = \sum_{\alpha=1}^{N} c_\alpha \mathbb{C}_\alpha,$$

$$\left(\bar{\mathbb{C}}^R\right)^{-1} = \bar{\mathbb{S}}^R = \langle \mathbb{S}(\boldsymbol{x}) \rangle = \frac{1}{V} \int_V \mathbb{S}(\boldsymbol{x}) \, \mathrm{d}V = \int_Q f(\boldsymbol{Q}) \mathbb{S}(\boldsymbol{Q}) \, \mathrm{d}Q = \sum_{\alpha=1}^{N} c_\alpha \mathbb{S}_\alpha, \qquad (5.58)$$

$$\bar{\mathbb{C}}^R \leq \bar{\mathbb{C}} \leq \bar{\mathbb{C}}^V;$$

the latter equation being meant in an energetic sense in a quadratic form.

For single-phase polycrystals of grains possessing cubic symmetry (cf. Eq. (2.26))

$$\mathbb{C} = \lambda_1 \mathbb{P}_1^C + \lambda_2 \mathbb{P}_2^C + \lambda_3 \mathbb{P}_3^C = (\lambda_1 - \lambda_2 + \lambda_3)\mathbb{P}_1^I + \lambda_3 \mathbb{P}_2^I + (\lambda_2 - \lambda_3)\boldsymbol{Q} \star \mathbb{D}_0, \quad (5.59)$$

the Voigt and Reuss bounds can be written using a split into an isotropic and anisotropic contribution (Böhlke and Bertram, 2001b), which yields for the effective stiffness according to Voigt

$$\begin{aligned}
\bar{\mathbb{C}}^V &= \int_Q f(\boldsymbol{Q}) \mathbb{C}(\boldsymbol{Q}) \, \mathrm{d}Q \\
&= (\lambda_1 - \lambda_2 + \lambda_3)\mathbb{P}_1^I + \lambda_3 \mathbb{P}_2^I + (\lambda_2 - \lambda_3) \int_Q f(\boldsymbol{Q})\boldsymbol{Q} \star \mathbb{D}_0 \, \mathrm{d}Q \\
&= \mathbb{C}^{VI} + (\lambda_2 - \lambda_3)\left(-\mathbb{P}_1^I - \frac{2}{5}\mathbb{P}_2^I + \int_Q f(\boldsymbol{Q})\boldsymbol{Q} \star \mathbb{D}_0 \, \mathrm{d}Q \right) \\
&= \mathbb{C}^{VI} + \underbrace{\sqrt{\frac{6}{5}}(\lambda_2 - \lambda_3)}_{\zeta^V} \underbrace{\frac{\sqrt{30}}{30}\left(-5\mathbb{P}_1^I - 2\mathbb{P}_2^I + 5\int_Q f(\boldsymbol{Q})\boldsymbol{Q} \star \mathbb{D}_0 \, \mathrm{d}Q \right)}_{\mathbb{V}' = \int_{SO(3)} f(\boldsymbol{Q})\boldsymbol{Q} \star \mathbb{V}'_c \, \mathrm{d}Q} \\
&= \mathbb{C}^{VI} + \zeta^V \mathbb{V}',
\end{aligned} \qquad (5.60)$$

and for the effective compliance tensor according to Sachs

$$\begin{aligned}
\bar{\mathbb{S}}^R &= \int_Q f(\boldsymbol{Q}) \mathbb{S}(\boldsymbol{Q}) \, \mathrm{d}Q \\
&= \mathbb{S}^{RI} + \sqrt{\frac{6}{5}}\left(\lambda_2^{-1} - \lambda_3^{-1} \right)\frac{\sqrt{30}}{30}\left(-5\mathbb{P}_1^I - 2\mathbb{P}_2^I + 5\int_Q f(\boldsymbol{Q})\boldsymbol{Q} \star \mathbb{D}_0 \, \mathrm{d}Q \right) \\
&= \mathbb{S}^{RI} + \zeta^R \mathbb{V}'.
\end{aligned} \qquad (5.61)$$

The anisotropic part of the elementary bounds is linear in the texture coefficient $\mathbb{V}'$ and contains two multiplicatively coupled parts. The scalar quantities ζ^V and ζ^R stem from the constitutive anisotropy of the grains, the other from the anisotropy of the orientation distribution function in terms of the fourth-order texture coefficient $\mathbb{V}'$, with $\|\mathbb{V}'\| \in [0, 1]$. The elementary bounds of cubic crystal aggregates are only isotropic if the fourth-order texture coefficient vanishes.

5.3.2 Second-order bounds

The elementary bounds are known to give poor results in terms of a broad spectrum of potential material behavior for materials with significant anisotropies on the microscale. Hashin and Shtrikman (1962) proposed a variational principle to get tighter bounds. In their method, they do not relate total stress and strain fields, but quantities that represent a deviation from a reference solution. By this choice, errors due to certain assumptions in the method influence much less the solution. Hashin and Shtrikman used the stress polarization $p(x) = (\mathbb{C}(x) - \mathbb{C}_0)[\varepsilon(x)] = \delta\mathbb{C}(x)[\varepsilon(x)]$, when $\mathbb{C}_0$ is a homogeneous reference stiffness, as such a quantity.

The following derivation is based on the works of Hashin and Shtrikman (1962, 1963); Walpole (1966a,b); Willis (1977, 1981).[10]

With the aforementioned stress polarization, the stress tensor takes the form

$$\sigma(x) = \mathbb{C}_0[\varepsilon(x)] + p(x). \tag{5.62}$$

If the polarizations $p(x)$ would be known, the strains solving the boundary value problem

$$\mathrm{div}\,(\sigma(x)) = \mathrm{div}\,(\mathbb{C}_0[\varepsilon(x)]) + \mathrm{div}\,(p(x)) = o \tag{5.63}$$

would be given by

$$\varepsilon(x) = \varepsilon_0 - \mathcal{G}(x)[p(x)], \tag{5.64}$$

with the reference strain ε_0 and the integral operator

$$\mathcal{G}(x)[p(x)] = \int_{V'} \Gamma(x - x')[p(x')]\,\mathrm{d}V'. \tag{5.65}$$

Γ includes the second derivatives of infinite body's Green's function[11] (cf. Eq. (5.26)) and it has the property $\mathcal{G}(x)[p(x)] = \mathcal{G}(x)[p(x) - \langle p(x) \rangle]$, which helps

---

[10]Recently, Berryman (2011, 2012) proposed an approach simplifying the computation of the HS bounds.

[11]It is valid to use infinite body's Green's function also for polycrystalline aggregates, see, e.g., Kröner (1972); Dederichs and Zeller (1972).

to handle the singularity of $\Gamma(\boldsymbol{x}) = \mathcal{O}(\|\boldsymbol{x}\|^{-3})$ (Willis, 1977). As can be seen from Eq. (5.63), the problem involving an inhomogeneous material without body forces was transferred to a problem with a homogeneous material and an apparent body force density $\boldsymbol{b}(\boldsymbol{x}) = \text{div}\,(\boldsymbol{p}(\boldsymbol{x}))$.

The polarization fulfilling the boundary value problem is given as solution of

$$\delta\mathbb{C}^{-1}(\boldsymbol{x})[\boldsymbol{p}(\boldsymbol{x})] + \mathcal{G}(\boldsymbol{x})[\boldsymbol{p}(\boldsymbol{x})] = \boldsymbol{\varepsilon}_0, \tag{5.66}$$

which is equivalent to the variational principle

$$\delta\mathcal{F} = \delta\left(\frac{1}{2}\boldsymbol{p}(\boldsymbol{x}) \cdot \delta\mathbb{C}^{-1}(\boldsymbol{x})[\boldsymbol{p}(\boldsymbol{x})] + \frac{1}{2}\boldsymbol{p}(\boldsymbol{x}) \cdot \mathcal{G}(\boldsymbol{x})[\boldsymbol{p}(\boldsymbol{x})] - \boldsymbol{p}(\boldsymbol{x}) \cdot \boldsymbol{\varepsilon}_0\right) \tag{5.67}$$

$$= \left(\delta\mathbb{C}^{-1}[\boldsymbol{p}(\boldsymbol{x})] + \mathcal{G}(\boldsymbol{x})[\boldsymbol{p}(\boldsymbol{x})] - \boldsymbol{\varepsilon}_0\right) \cdot \delta\boldsymbol{p}(\boldsymbol{x}). \tag{5.68}$$

When considering a heterogeneous material with several inclusions/inhomogeneities, the volume average of Eq. (5.67) needs to reach stationarity $\delta\bar{\mathcal{F}} = \langle\delta\mathcal{F}\rangle = 0$, where it can be shown that $\bar{\mathcal{F}} = W_0 - \bar{W}$ (see, e.g., Willis (1977)). W_0 is the strain energy of the homogeneous comparison material and $\bar{W} = \langle W\rangle$ the effective strain energy of the heterogeneous material. The second variation $\delta^2\bar{\mathcal{F}}$ delivers information on $\bar{\mathcal{F}}$ being a minimum or maximum

$$\delta^2\bar{\mathcal{F}} = \langle\delta\boldsymbol{p}(\boldsymbol{x}) \cdot (\delta\mathbb{C}^{-1}(\boldsymbol{x}) + \mathcal{G}(\boldsymbol{x}))[\delta\boldsymbol{p}(\boldsymbol{x})]\rangle \begin{cases} > 0 & \bar{\mathcal{F}}\ \text{min} \\ < 0 & \bar{\mathcal{F}}\ \text{max} \end{cases}, \tag{5.69}$$

which in fact is a quadratic form. Willis (1977) examined the operator appearing in Eq. (5.69) and showed that if $\delta\mathbb{C}$ is positive definite, also $\delta\mathbb{C}^{-1}(\boldsymbol{x}) + \mathcal{G}(\boldsymbol{x})$ is positive definite and if $\delta\mathbb{C}$ is negative definite, also $\delta\mathbb{C}^{-1}(\boldsymbol{x}) + \mathcal{G}(\boldsymbol{x})$ is negative definite.

Considering the interpretation of the Hashin-Shtrikman functional $\bar{\mathcal{F}} = W_0 - \bar{W}$ as the difference in strain energies of the comparison and effective material, it can be concluded that

$$\delta\mathbb{C}\begin{cases} \text{pos.def.,} & \delta^2\bar{\mathcal{F}} > 0, \quad \bar{W} = \bar{W}_{\text{max}} \\ \text{neg.def.,} & \delta^2\bar{\mathcal{F}} < 0, \quad \bar{W} = \bar{W}_{\text{min}} \end{cases}. \tag{5.70}$$

For finding the optimal bounds, Nadeau and Ferrari (2001) - based on the work of Kröner (1977) - proposed a generalized scheme being applicable for arbitrarily composed heterogeneous materials. The important result of their paper is the fact that the optimal bounds are found if $\mathbb{C}_0$ is isotropic. By using this method, one can show that in case of a polycrystal consisting of cubic single crystals with $\mathbb{C}(\boldsymbol{x}) = \boldsymbol{Q} \star \sum_{i=1}^{3} \lambda_i \mathbb{P}_i^C$, the optimal Hashin-Shtrikman lower bound is obtained for $\mathbb{C}_0^{HS\pm} = \lambda_1^{HS\pm} \mathbb{P}_1^I + \lambda_2^{HS\pm} \mathbb{P}_2^I$ with $\lambda_1^{HS-} = \lambda_1$ and $\lambda_2^{HS-} = \min(\lambda_2, \lambda_3)$, and equivalently, for the optimal upper bound the identical $\lambda_1^{HS+} = \lambda_1$ but $\lambda_2^{HS+} = \max(\lambda_2, \lambda_3)$. The effective stiffness in the generalized Hashin-Shtrikman scheme is given by

$$\bar{\mathbb{C}}^{HS\pm} = \langle \mathbb{C}\mathbb{A}^{HS\pm} \rangle, \tag{5.71}$$

where $\mathbb{A}^{HS\pm}$ is the strain localization tensor being derived from Eq. (5.64) with eliminating ε_0, so that

$$\varepsilon(\boldsymbol{x}) = (\mathbb{I}^S + \mathcal{G}\delta\mathbb{C})^{-1}[\varepsilon_0] = (\mathbb{I}^S + \mathcal{G}\delta\mathbb{C})^{-1} \langle (\mathbb{I}^S + \mathcal{G}\delta\mathbb{C})^{-1} \rangle^{-1} [\bar{\varepsilon}] = \mathbb{A}^{HS\pm}[\bar{\varepsilon}] \tag{5.72}$$

and $\delta\mathbb{C} = \delta\mathbb{C}^{HS\pm} = \mathbb{C}(\boldsymbol{x}) - \mathbb{C}_0^{HS\pm}$. Note that if $\varepsilon_0 = \bar{\varepsilon}$, the scaling factor $\langle (\mathbb{I}^S + \mathcal{G}\delta\mathbb{C})^{-1} \rangle = \mathbb{I}^S$. As discussed by Willis (1977, 1981), for piecewise constant polarizations $\boldsymbol{p}(\boldsymbol{x}) = \boldsymbol{p}_\alpha$, $\forall \boldsymbol{x} \in V_\alpha$, ellipsoidal two-point statistics and no long range order correlation in the heterogeneous material, $\mathcal{G}$ becomes $\mathbb{P}_0$ (see Eq. (5.51)), so that

$$\bar{\mathbb{C}}^{HS\pm} = \langle \mathbb{C}(\mathbb{I}^S + \mathbb{P}_0\delta\mathbb{C}^{HS\pm})^{-1} \langle (\mathbb{I}^S + \mathbb{P}_0\delta\mathbb{C}^{HS\pm})^{-1} \rangle^{-1} \rangle. \tag{5.73}$$

For the special case of effective isotropic properties of the polycrystal, the isotropic Hashin-Shtrikman bounds (Hashin and Shtrikman, 1962) can be explicitly given in terms of isotropic eigenvalues. The spherical eigenvalue of the HS bounds in case of cubic crystal symmetry equals the spherical eigenvalue of the single crystal $\bar{\lambda}_1^{HS\pm} = \lambda_1$. The second eigenvalue for the upper bound and the lower bound are given for the case that $\lambda_3 > \lambda_2$ by

$$\bar{\lambda}_2^{HS+} = \lambda_3 + 2\left(\frac{5}{\lambda_2 - \lambda_3} + 6\frac{\lambda_1 + 3\lambda_3}{5\lambda_3(\lambda_1 + 2\lambda_3)}\right)^{-1}, \tag{5.74}$$

$$\bar{\lambda}_2^{HS-} = \lambda_2 + 3\left(\frac{5}{\lambda_3 - \lambda_2} + 4\frac{\lambda_1 + 3\lambda_2}{5\lambda_2(\lambda_1 + 2\lambda_2)}\right)^{-1}. \tag{5.75}$$

5.3.3 Higher-order bounds

The more information on the statistical distribution of the phases/grains of the heterogeneous material is available, the more precise the real material behavior can be predicted. For the bounding techniques this means that increasing the number n of n-point correlation functions, which deliver higher-order statistical information on the microstructure, being included in the theory, the closer the upper and lower bounds to each other. In this sense, Dederichs and Zeller (1973)[12] introduced higher-order bounds of odd order and Kröner (1977) extended the theory to higher-order bounds of even order. In general, they use the expansion of the effective stiffness into a Neumann series, which is common in perturbation theory (see also, e.g., Beran and McCoy, 1970; Morawiec, 2004)

$$
\begin{aligned}
\bar{\mathbb{C}} &= \langle \mathbb{C}(\mathbb{I}^S + \mathcal{G}\delta\mathbb{C})^{-1}\rangle \langle (\mathbb{I}^S + \mathcal{G}\delta\mathbb{C})^{-1}\rangle^{-1} \\
&= \mathbb{C}_0 + \langle \delta\mathbb{C}(\mathbb{I}^S + \mathcal{G}\delta\mathbb{C})^{-1}\rangle \langle (\mathbb{I}^S + \mathcal{G}\delta\mathbb{C})^{-1}\rangle^{-1} \\
&= \mathbb{C}_0 + \langle \delta\mathbb{C}\rangle - \langle \delta\mathbb{C}\mathcal{G}\delta\mathbb{C}\rangle + \dots,
\end{aligned}
\tag{5.76}
$$

or, alternatively, the smoothed or renormalized version

$$
\bar{\mathbb{C}} = \mathbb{C}_0 + \langle \delta\mathbb{C}\rangle - \langle \delta\mathbb{C}\mathcal{G}(\delta\mathbb{C} - \langle \delta\mathbb{C}\rangle)\rangle + \dots,
\tag{5.77}
$$

which is truncated after a certain term. Using such a perturbation expansion in terms of compliances and the associated operator $\mathcal{H} = \mathbb{C}_0 - \mathbb{C}_0\mathcal{G}\mathbb{C}_0$, accordingly generates a lower bound.

By the concept of disorder, Kröner (1977) explicitly gives second- and third-order bounds for material classes, which are so-called 'overall grade 2' and 'overall grade 3', respectively and shows that for totally disordered materials, his bounds conceptually include the Hashin-Shtrikman bounds (Hashin and Shtrikman, 1963) in the second-order case and give narrower bounds for the third-order case.

Willis (1981), being inspired by Kröner (1977), proposed to use improved higher-order bounds, so not just truncating the Neumann series after a certain term (like Dederichs and Zeller (1973)), but using trial fields and minimizing the energy to find the optimum. Actually, Willis procedure is similar to Walpole's bounds (Walpole, 1966a,b), but without the restriction to isotropic two-point statistics.

---

[12]See also Zeller and Dederichs (1973) for higher order approximations (not bounds) being called the $\mathcal{T}$-matrix formalism, which is based on a method from quantum mechanical scattering theory.

5.4 Singular approximation

5.4.1 Singular approximation of effective properties

The singular approximation (Fokin (1972, 1973); Morawiec (2004); Böhlke et al. (2010)[13]) is also based on Green's function and a comparison material with stiffness $\mathbb{C}_0$ which is a free parameter, so that the local strain field (cf. Eq. (5.64)) in the heterogeneous material can be expressed by

$$\varepsilon(\boldsymbol{x}) = \varepsilon_0 - \mathcal{G}\delta\mathbb{C}[\varepsilon(\boldsymbol{x})], \tag{5.78}$$

with the integral operator $\mathcal{G}$ (see Eq. (5.65)). The general property of the fourth-order tensor Γ in the integral operator (Dederichs and Zeller, 1973; Torquato, 2002; Morawiec, 2004) is that it can be decomposed into a singular and a nonlocal part $\Gamma(\boldsymbol{r}) = \Gamma_0\delta(\boldsymbol{r}) + \Gamma_1(\boldsymbol{r})$ with r the distance between two points $\boldsymbol{x}$ and $\boldsymbol{x}'$ given by $r = \boldsymbol{x} - \boldsymbol{x}'$ and the Dirac distribution $\delta(\boldsymbol{r})$. Γ_0 is a constant tensor and the nonlocal part Γ_1 has the property $\Gamma_1(\alpha\boldsymbol{r}) = \alpha^{-3}\Gamma_1(\boldsymbol{r})$, but the integral operator is invariant under the change $\boldsymbol{r} \to \alpha\boldsymbol{r}$.

The singular approximation of Γ is obtained by neglecting the nonlocal part (Fokin, 1972, 1973)

$$\Gamma(\boldsymbol{r}) \approx \Gamma_0\delta(\boldsymbol{r}), \tag{5.79}$$

so that after eliminating the comparison strain ε_0, the strain localization relation reads

$$\varepsilon = \mathbb{A}^{SA}[\bar{\varepsilon}], \qquad \mathbb{A}^{SA} = (\mathbb{I}^S + \Gamma_0\delta\mathbb{C})^{-1}\langle(\mathbb{I}^S + \Gamma_0\delta\mathbb{C})^{-1}\rangle^{-1}. \tag{5.80}$$

The effective stiffness tensor is given by

$$\bar{\mathbb{C}}^{SA} = \langle\mathbb{C}\mathbb{A}^{SA}\rangle = \langle\mathbb{C}(\mathbb{I}^S + \Gamma_0\delta\mathbb{C})^{-1}\rangle\langle(\mathbb{I}^S + \Gamma_0\delta\mathbb{C})^{-1}\rangle^{-1} = \langle\mathbb{C}\mathbb{M}\rangle\langle\mathbb{M}\rangle^{-1}, \tag{5.81}$$

or equivalently, the effective compliance tensor being reciprocal to the effective stiffness

$$\bar{\mathbb{S}}^{SA} = \langle\mathbb{S}(\mathbb{I}^S + \Lambda_0\delta\mathbb{S})^{-1}\rangle\langle(\mathbb{I}^S + \Lambda_0\delta\mathbb{S})^{-1}\rangle^{-1} = \langle\mathbb{S}\mathbb{N}\rangle\langle\mathbb{N}\rangle^{-1}, \tag{5.82}$$

with $\Lambda_0 = \mathbb{C}_0 - \mathbb{C}_0\Gamma_0\mathbb{C}_0$, $\delta\mathbb{S} = \mathbb{S} - \mathbb{S}_0$ and $\mathbb{S}_0 = \mathbb{C}_0^{-1}$.

The singular approximation fulfills the condition that the effective stiffness and compliance are reciprocal $\bar{\mathbb{S}}^{SA} = (\bar{\mathbb{C}}^{SA})^{-1}$. With $\mathbb{C}\mathbb{M} = \mathbb{N}\mathbb{C}_0$ and $\mathbb{S}\mathbb{N} = \mathbb{M}\mathbb{S}_0$, this

---

[13] The derivations in section 5.4 closely follow the work of Böhlke et al. (2010).

can easily be proved, so that the effective elasticity tensors are equivalently given by

$$\bar{\mathbb{C}}^{SA} = \langle \mathbb{N} \rangle \mathbb{C}_0 \langle \mathbb{M} \rangle^{-1} = \left(\langle \mathbb{M} \rangle \mathbb{S}_0 \langle \mathbb{N} \rangle^{-1} \right)^{-1} = \left(\bar{\mathbb{S}}^{SA} \right)^{-1}. \tag{5.83}$$

Since the singular approximation neglects the nonlocal part of the integral operator, morphology aspects of the phases are not accounted for, so that, e.g., for polycrystalline aggregates, the method inherently only incorporates anisotropy effects due to crystallographic texture. Hence, Γ_0 is equivalent to Hill's polarization tensor $\mathbb{P}_0$ when evaluated for isotropic two-point statistics.

5.4.2 Singular approximation in terms of texture coefficients

For cubic crystal aggregates, with stiffnesses of the crystals being given by $\mathbb{C}_\gamma = \boldsymbol{Q}_\gamma \star \sum_{i=1}^{3} \lambda_i \mathbb{P}_i^C$ and choosing an isotropic comparison medium $\mathbb{C}_0 = \lambda_1^0 \mathbb{P}_1^I + \lambda_2^0 \mathbb{P}_2^I$, it is possible to write

$$\delta\mathbb{C} = \sum_{i=1}^{3} (\lambda_i - \lambda_i^0) \mathbb{P}_i^C, \tag{5.84}$$

since $\mathbb{P}_2^I = \mathbb{P}_2^C + \mathbb{P}_3^C$ and defining $\lambda_2^0 = \lambda_3^0$. Similarly, Γ_0 can be given by $\Gamma_0 = g_1 \mathbb{P}_1^I + g_2 \mathbb{P}_2^I = g_1 \mathbb{P}_1^C + g_2 \mathbb{P}_2^C + g_3 \mathbb{P}_2^C$ with $g_2 = g_3$. The constants g_1 and g_2 are implicitly given by Eq. (5.57). When using the projector properties being idempotence $\mathbb{P}_\alpha^C \mathbb{P}_\alpha^C = \mathbb{P}_\alpha^C$, biorthogonality $\mathbb{P}_\alpha^C \mathbb{P}_\beta^C = \mathbb{O}$ ($\alpha \neq \beta$), and completeness $\sum_{\alpha=1}^{3} \mathbb{P}_\alpha^C = \mathbb{I}^S$ (these characteristics equivalently hold for the two isotropic projectors), the fourth-order tensor $\mathbb{M}$ in Eq. (5.81) is given by

$$\mathbb{M} = \left(\mathbb{I}^S + \Gamma_0 \delta\mathbb{C} \right)^{-1} = \sum_{\alpha=1}^{3} m_\alpha \mathbb{P}_\alpha^C, \qquad m_\alpha = \left(1 + g_\alpha (\lambda_\alpha - \lambda_\alpha^0) \right)^{-1}, \tag{5.85}$$

and the other term becomes

$$\mathbb{C}\mathbb{M} = \sum_{\alpha=1}^{3} \lambda_\alpha m_\alpha \mathbb{P}_\alpha^C. \tag{5.86}$$

Since an elasticity tensor with cubic crystal symmetry can also be given in terms of a fourth-order texture coefficient $\mathbb{V}_C'$ (see Eqs. (5.59), (5.60) and (5.61))

$$\mathbb{C} = \sum_{\alpha=1}^{3} \lambda_\alpha \mathbb{P}_\alpha^C = \lambda_1 \mathbb{P}_1^I + \left(\frac{2}{5}\lambda_2 + \frac{3}{5}\lambda_3 \right) \mathbb{P}_2^I + \sqrt{\frac{6}{5}}(\lambda_2 - \lambda_3) \mathbb{V}_C', \tag{5.87}$$

one concludes

$$\mathbb{M} = m_1 \mathbb{P}_1^I + \left(\frac{2}{5}m_2 + \frac{3}{5}m_3\right)\mathbb{P}_2^I + \sqrt{\frac{6}{5}}(m_2 - m_3)\mathbb{V}_C', \tag{5.88}$$

$$\mathbb{CM} = \lambda_1 m_1 \mathbb{P}_1^I + \left(\frac{2}{5}\lambda_2 m_2 + \frac{3}{5}\lambda_3 m_3\right)\mathbb{P}_2^I + \sqrt{\frac{6}{5}}(\lambda_2 m_2 - \lambda_3 m_3)\mathbb{V}_C', \tag{5.89}$$

the volume averages of which are quite similar by only substituting $\mathbb{V}_C'$ by $\mathbb{V}' = \langle \mathbb{V}_C' \rangle$.

Thus, the following representation of the effective stiffness tensor is derived

$$\bar{\mathbb{C}}^{SA} = (3K\mathbb{P}_1^I + \alpha\mathbb{P}_2^I + \beta\mathbb{V}')(\mathbb{P}_1^I + \tilde{\alpha}\mathbb{P}_2^I + \tilde{\beta}\mathbb{V}')^{-1}, \tag{5.90}$$

where $\lambda_1 = 3K$ is the exact bulk modulus of the aggregate, so that it is obvious to set $\lambda_1^0 = \lambda_1$. The parameters α, β, $\tilde{\alpha}$ and $\tilde{\beta}$ are determined by

$$\alpha = \frac{2}{5}\lambda_2 m_2 + \frac{3}{5}\lambda_3 m_3, \ \beta = \sqrt{\frac{6}{5}}(\lambda_2 m_2 - \lambda_3 m_3), \ \tilde{\alpha} = \frac{2}{5}m_2 + \frac{3}{5}m_3, \ \tilde{\beta} = \sqrt{\frac{6}{5}}(m_2 - m_3). \tag{5.91}$$

Similarly, the effective compliance tensor is given by

$$\bar{\mathbb{S}}^{SA} = ((3K)^{-1}\mathbb{P}_1^I + \zeta\mathbb{P}_2^I + \eta\mathbb{V}')(\mathbb{P}_1^I + \tilde{\zeta}\mathbb{P}_2^I + \tilde{\eta}\mathbb{V}')^{-1}, \tag{5.92}$$

with

$$\zeta = \frac{2}{5}\lambda_2^{-1}n_2 + \frac{3}{5}\lambda_3^{-1}n_3, \ \eta = \sqrt{\frac{6}{5}}(\lambda_2^{-1}n_2 - \lambda_3^{-1}n_3), \ \tilde{\zeta} = \frac{2}{5}n_2 + \frac{3}{5}n_3, \ \tilde{\eta} = \sqrt{\frac{6}{5}}(n_2 - n_3), \tag{5.93}$$

and $n_\alpha = (1 + h_\alpha(\lambda_\alpha - \lambda_\alpha^0))^{-1}, \alpha = 2,3$ as well as $\mathbf{\Lambda}_0 = \mathbb{C}_0 - \mathbb{C}_0\mathbf{\Gamma}_0\mathbb{C}_0 = h_1\mathbb{P}_1^I + h_2\mathbb{P}_2^I$ and $h_2 = h_3$.

If the orientation distribution is isotropic, then $f(\mathbf{Q}) = 1$ and $\mathbb{V}' = \mathbb{O}$ hold. As a result, the singular approximation simplifies to

$$\bar{\mathbb{C}}^{SA} = 3K\mathbb{P}_1^I + \frac{\alpha}{\tilde{\alpha}}\mathbb{P}_2^I, \qquad \bar{\mathbb{S}}^{SA} = \frac{1}{3K}\mathbb{P}_1^I + \frac{\zeta}{\tilde{\zeta}}\mathbb{P}_2^I. \tag{5.94}$$

An inspection of these equations shows (see also Morawiec (2004)) that the Voigt and Reuss bounds are obtained for $\lambda_2^0 \to \infty$ and $\lambda_2^0 \to 0$, respectively. The upper and the lower Hashin-Shtrikman bound follow for $\lambda_2^0 = \max(\lambda_2, \lambda_3)$ and $\lambda_2^0 = \min(\lambda_2, \lambda_3)$, respectively. This is in contrast to the approach by Huang and Man (2008), where a bounding technique is suggested that includes the lower Hashin-Shtrikman bound, however, not exactly the upper Hashin-Shtrikman bound for the special case of an isotropic orientation distribution.

5.5 Self-consistent scheme

In the self-consistent sense, every constituent of the heterogeneous material is embedded as a single inclusion (using Eshelbys solution) into a matrix material, which has the effective properties of the surrounding material (Hershey, 1954; Kröner, 1958; Hill, 1965b; Budiansky, 1965)[14]. Due to this particular choice of matrix material, the method has an inherently implicit character. By the introduction of a homogeneous reference medium with stiffness tensor equal to the effective medium stiffness $\mathbb{C}_0 = \bar{\mathbb{C}}$, and under the assumptions of piecewise constant stress polarizations as well as statistical homogeneity, no long-range order correlation and ellipsoidal two-point statistics, the effective material behavior can be similarly derived as the Hashin-Shtrikman bounds (section 5.3.2), so that the effective stiffness is given by

$$\bar{\mathbb{C}}^{SC} = \langle \mathbb{C}\mathbb{A}^{SC} \rangle = \langle \mathbb{C}(\mathbb{I}^S + \mathbb{P}_0(\bar{\mathbb{C}}^{SC})\delta\mathbb{C}(\bar{\mathbb{C}}^{SC}))^{-1} \rangle \langle (\mathbb{I}^S + \mathbb{P}_0(\bar{\mathbb{C}}^{SC})\delta\mathbb{C}(\bar{\mathbb{C}}^{SC}))^{-1} \rangle^{-1},$$

$$(5.95)$$

which is exactly Eq. (5.73) with $\mathbb{C}_0 = \bar{\mathbb{C}}$. The estimate (Eq. (5.95)) is also self-consistent in the sense that it delivers the same result irrespective of whether stiffnesses or compliances are considered for the averaging scheme. It should be noted that in general $\bar{\mathbb{C}}$ is anisotropic. Therefore, the polarization tensor $\mathbb{P}_0$ is also anisotropic and has to be evaluated numerically. The numerical integration of $\mathbb{P}_0$ and the necessity of iterative evaluation of the effective stiffness render the self-consistent scheme computationally more expensive compared to the singular approximation. Nevertheless, it is more general due to the incorporation of morphological anisotropies in the ellipsoidal sense in $\mathbb{P}_0$ and, furthermore, the free parameter $\mathbb{C}_0$ has been eliminated in a physically obvious way.

5.6 Computational homogenization based on full-field simulations

In contrast to the aforementioned semi-analytical models which relate phase averages of stress and strains and, thereby, allow to determine the effective stiffness of the considered representative volume element (RVE), finite element

---

[14]Note also the behavior of the SC method in the limiting cases of voids, rigid particles, incompressible phases, etc., discussed by Budiansky (1965); Milton (2002).

computations delivering full fields of stresses and strains can be used to determine the effective material behavior. In this case, the microstructure of the RVE is entirely spatially resolved and discretized with finite elements.

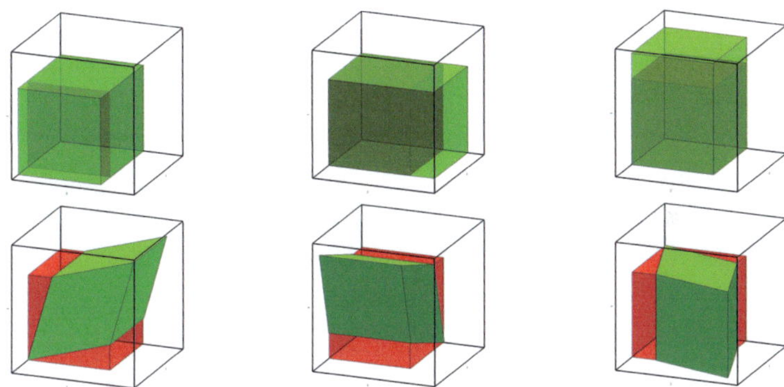

Figure 5.3: Six orthogonal load cases to determine the effective stiffness of a RVE (red: undeformed, green: deformed)

Six orthogonal load cases $\varepsilon_\eta = \varepsilon_0 \boldsymbol{B}_\eta$ $(\eta = 1, \ldots, 6)$ describing linear elastic boundary value problems, where $\boldsymbol{B}_\eta$ is the orthonormal basis of symmetric second-order tensors (see, e.g., Cowin (1989) - also known as modified Voigt notation), have to be solved (Fig. 5.3). Each of them provides one column of the effective stiffness $\bar{\mathbb{C}}^{FE}$ by dividing the obtained volume averaged stress tensor by the amplitude ε_0

$$\bar{\mathbb{C}}^{FE}[\boldsymbol{B}_\eta] = \frac{1}{\varepsilon_0} \langle \boldsymbol{\sigma}(\varepsilon_\eta) \rangle. \tag{5.96}$$

Full-field computations of the effective material response are computationally significantly more expensive than the usage of mean-field approaches. However, the solution is very precise due to the detailed resolution of the grain interactions. This is pretty meaningful and valuable if only one specific microstructure is to be investigated.

5.7 Applications

In this section, two problems involving cubic crystal aggregates are considered, which highlight two completely different sources of effective anisotropy. On the one hand, small tensile specimen are investigated. In this case, the orientation distribution of the grains is assumed to be uniform. Due to the small dimensions of the samples, which is in the order of magnitude as the grain size, the macroscopically observable anisotropy is a resultant of the small number of grains forming the whole specimen. On the other hand, the elasticity of rolled sheets is evaluated using different approaches presented in this section. In that case, the effective anisotropy results from the rolling process, in which, contrarily to the microspecimen investigations, crystallographic as well as morphological anisotropies play a crucial role.

5.7.1 Anisotropy of oligocrystals

In many applications, nowadays, microcomponents are effectively employed. These small parts, having at least two dimensions in the submillimeter range (Geiger et al., 2001), are of increasing importance in, e.g., micro-system technologies or micro-electromechanical systems (Ko, 2007). For a save prediction of the structural behavior of microparts, the material heterogeneities have to be studied intensively. The small dimensions of the parts result in only a small number of grains making up the whole component (oligocrystalline aggregate), so that the real specimen is smaller than a RVE. In this case, the concept of effective properties, which are insensitive to microstructural changes maintaining the statistics, fails, so that apparent elastic properties have to be analyzed (Huet, 1990; Ostoja-Starzewski, 1993; Hazanov and Huet, 1994).

The investigations in this section are motivated by research on microcomponents made of Stabilor®G that were produced and tested within the scope of the SFB 499 at the Karlsruhe Institute of Technology (KIT) (e.g., Baumeister et al., 2004; Auhorn et al., 2006). Stabilor®G is an alloy mainly consisting of gold, the further constituents being silver, copper, palladium and zinc. The investigated tensile specimen have a gauge length of 0.78 mm and a cross sectional area of 0.26×0.13 mm^2 (Fig. 5.4(a)). One special set of casting conditions is considered in this work resulting in an average grain size of 35.1 μm (Fig. 5.4(b)), so that the tensile specimen in total consists of approximately 600 grains. Although

(a)   (b)   (c)

Figure 5.4: (a) Casted micro tensile specimens with gate, (b) microstructure, and (c) experimental stress strain curves of Stabilor®G (figures taken from Auhorn, 2005)

the manufacturing process does not cause any preferential orientations in the specimen, the macroscopic behavior strongly depends on the particular orientation distribution in the specimen due to the small number of grains present in the sample. The scattering of the plastic behavior is much more pronounced than the one of elastic properties (Fig. 5.4(c)), but the latter is, nevertheless, significant. The experimentally determined Young's modulus shows to be $E = 103 \pm 2\text{GPa}$. The following paragraphs mainly summarize results published in Böhlke et al. (2008); Jöchen and Böhlke (2009c); Böhlke et al. (2010).

Elastic behavior of Stabilor®G

In a first step, the finite element program ABAQUS has been used to carry out simulations of uniaxial tensile tests. In the finite element model, the microstructure of the specimen is represented by a periodic Voronoi tessellation. Since the volume of the microspecimen is fixed, the Poisson distribution of the generator points is equal to a uniform distribution of generator points in the specimen. The production process and the experiments do not give any indication for a crystallographic texture, so that the crystal orientations of the grains are assumed to be equi-probable. An isotropic discrete orientation distribution can be generated based on three random numbers $x_i \in [0, 1]$, $i = 1, 2, 3$, transforming into Euler angles (Bunge, 1993) by

$$\varphi_1 = 2\pi x_1, \quad \Phi = \text{acos}(-1 + 2x_2), \quad \varphi_2 = 2\pi x_3. \tag{5.97}$$

The tensile specimen has been discretized on the one hand by hexahedral elements using a structured mesh ($100 \times 30 \times 13$ C3D8 elements), and on the other hand by tetrahedral elements (C3D10 elements) using an unstructured mesh. Comparative calculations with a structured and unstructured mesh (cf. Figure 3.4) with approximately the same number of degrees of freedom have been carried out in order to analyze the local stress fields, whereas Figure 5.5 shows the smearing at the grain boundaries.

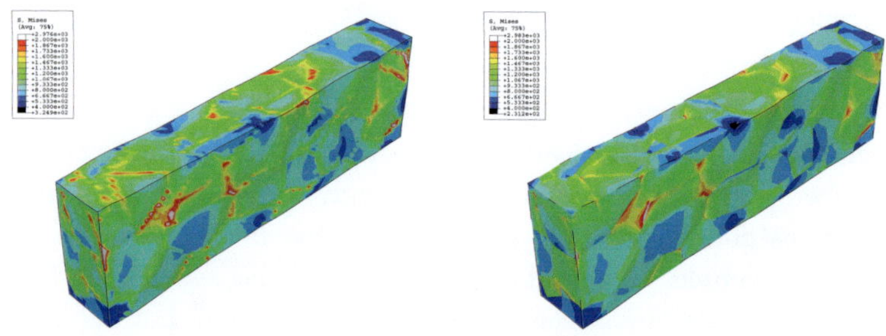

Figure 5.5: Stress distribution in structured (left) and unstructured mesh (right)

When analyzing the distribution of the von Mises stress, it shows, however, to agree well in both cases of different meshing strategies (Figure 5.6(a)), so that the usage of a structured mesh is admissible. An extensive investigation of the influence of the mesh type has been done in (Fritzen et al., 2009). In the scope of this work, the structured mesh is chosen for simplicity, since an ensemble averaging will be performed over several hundred microspecimens with different microstructures. Based on the approach suggested by Ostoja-Starzewski (2008), scale-dependent upper and lower bounds of the elastic strain energy density could be derived based on pure displacement or traction boundary conditions, respectively. In this study, mixed boundary conditions are applied, meaning that except for the tensile direction, where the displacement is prescribed, on all the other boundaries traction boundary conditions are applied (traction free boundaries).

For the estimation of the elastic constants of Stabilor®G, at first, the elastic behavior of polycrystals made of gold are examined with the following elastic

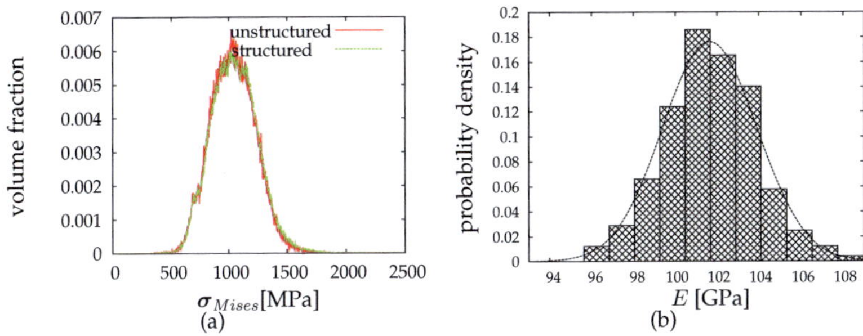

Figure 5.6: (a) Volume fraction of local von Mises stress in structured and un-structured mesh ; (b) numerically determined distribution of Young's modulus compared to Gaussian distribution

constants: $C_{1111} = 185$ GPa, $C_{1122} = 158$ GPa and $C_{1212} = 39.7$ GPa (Simmons and Wang, 1971). Based on the ensemble average over 300 simulations, each having a different microstructure with 600 grains, a factor of approximately 1.4 has been found for a uniform scaling of the elastic constants of gold so as to reproduce the experimentally observed mean value $E^{exp} = 103$ GPa of Stabilor®G. Hence, the single crystalline constants of Stabilor®G are assumed to be $C_{1111} = 255.3$ GPa, $C_{1122} = 218$ GPa and $C_{1212} = 54.8$ GPa.

Analogously, 300 different microstructures have been examined numerically and evaluated statistically for different discrete values for the numbers of grains in the specimen, in order to identify the elastic behavior of Stabilor®G. The grain size distribution in this case follows the one for Poisson-Voronoi tessellations, the distribution of relative volumes being exemplarily given by Kumar et al. (1992) for 300 Voronoi cells by $c = 1 \pm 0.42$.

Approximately, the determined distribution of Young's modulus is of Gaussian type, so that its distribution is completely described by a mean value and a standard deviation. Representative for each of the investigated numbers of grains, Figure 5.6(b) shows the distribution of Young's modulus of Stabilor®G determined by simulations for 200 grains in the specimen, compared to the Gaussian bell-shaped curve.

In Figure 5.7, the results of the aforementioned FE simulations are depicted, which are, more precisely, the mean value as well as the standard deviation of

Young's modulus with respect to the number of grains in the specimen. By gradually increasing the number of grains in the model, the elastic material behavior gets less anisotropic, which is indicated by a decreasing scattering. Moreover, the analytically determined isotropic first and second-order bounds are shown, whereas, apparently, the mean value of Young's modulus determined by the simulations tends between the Hashin-Shtrikman bounds, being closer to the upper one for larger numbers of grains.

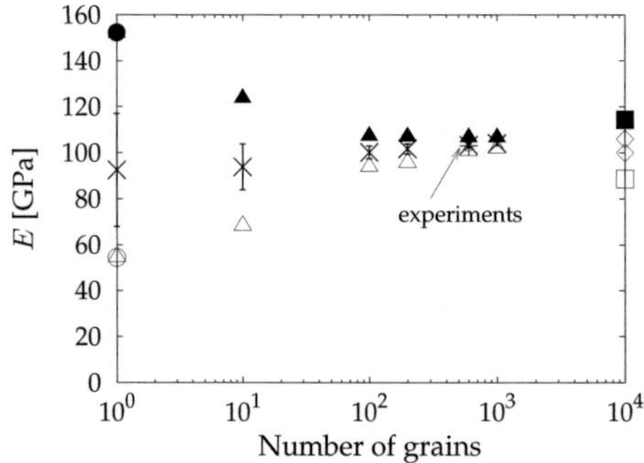

Figure 5.7: Young's modulus of Stabilor®G for different numbers of grains in the sample: ■ Voigt upper bound, □ Reuss lower bound, ◇ Hashin-Shtrikman bounds, × harmonic mean of Young's modulus from FE simulations, ▲ maximum Young's modulus in FE simulations, △ minimum Young's modulus in FE simulations, • Young's modulus in ⟨111⟩-direction, ○ Young's modulus in ⟨100⟩-direction, + experimental results

In a second step, analytical homogenization methods are applied, where at first the isotropic bounds and the isotropic version of the singular approximation (cf. section 5.4) are evaluated. The singular approximation (SA) depends on the chosen reference material $\mathbb{C}_0$. Since the effective value of the bulk modulus is known, $\lambda_1^0 = 3K = \lambda_1$ is used. The geometric mean gives a unique estimate for the effective stiffness, being the inverse of the effective compliance $\mathbb{C}^{G-1} = \mathbb{S}^G$. The geometric mean of the local elasticity tensors is defined by

$$\mathbb{C}^G = \exp\left(\int_Q f(\boldsymbol{Q}) \ln\left(\mathbb{C}(\boldsymbol{Q})\right) dQ\right), \qquad \mathbb{S}^G = \exp\left(\int_Q f(\boldsymbol{Q}) \ln\left(\mathbb{S}(\boldsymbol{Q})\right) dQ\right) \quad (5.98)$$

(Aleksandrov and Aisenberg, 1966). It can be shown that for cubic crystal aggregates, the geometric mean is bounded by the aforementioned elementary bounds (Böhlke and Bertram, 2001a). In the isotropic case, the geometric mean of the local stiffness is given by $\mathbb{C}^{GI} = 3K^G \mathbb{P}_1^I + 2G^G \mathbb{P}_2^I$ with $3K^G = \lambda_1$ and $2G^G = \lambda_2^{\frac{2}{5}} \lambda_3^{\frac{3}{5}}$. Therefore, it seems to be most consistent to choose the geometric mean to define the reference material (SA-G: $\lambda_2^0 = 2G^G$). To show the influence of the choice of reference material, the Voigt and the Reuss average are additionally used to specify the reference material (SA-V: $\lambda_2^0 = 2G^V$, SA-R: $\lambda_2^0 = 2G^R$). Independent of the chosen reference material, the singular approximation is located between the Hashin-Shtrikman bounds. Furthermore, it agrees quite well with the experimentally determined mean value as well as with the one found by FE simulations, although in both cases the material is not absolutely isotropic. Numerical values for the first and second-order isotropic bounds as well as for the singular approximation with $\mathbb{V}' = \mathbb{O}$, with $\mathbb{O}$ the fourth-order zero tensor, are given in Table 5.1.

	Voigt	Reuss	HS+	HS-	SA-V	SA-G	SA-R	Exp.
E [GPa]	114.3	88.6	106.0	100.0	104.4	103.7	102.9	103

Table 5.1: Isotropic bounds and estimates for Stabilor®G

Now, the anisotropy is taken into account. In order to estimate the elastic properties based on the singular approximation, the following procedure is applied. 300 discrete orientation distributions have been generated. For each of the distributions, the texture coefficient $\mathbb{V}'$ is computed and used to evaluate the anisotropic version of the singular approximation (Eq. (5.92)) in terms of the compliance tensor. From the compliance tensor, the reciprocal of Young's modulus in direction n is computed with

$$\frac{1}{E(\boldsymbol{n})} = (\boldsymbol{n} \otimes \boldsymbol{n}) \cdot \mathbb{S}[\boldsymbol{n} \otimes \boldsymbol{n}] \tag{5.99}$$

and ensemble averaged. Furthermore, the anisotropic versions of the one-point bounds (Eqs. (5.60) and (5.61)) are determined. The corresponding results are depicted in Figure 5.8. In the case of 600 grains in the specimen, which is the reference value given by experiments, the proposed mean value does not differ significantly (0.1%) from the isotropic estimates given in Table 5.1. This also became apparent by the predictions based on FE simulations. On the other hand, for smaller numbers of grains, the FE solution for the mean value of Young's modulus shows to be softer than the singular approximation result. This can probably be explained by the shape of the specimen being longer in tensile direction than in the other directions. Therefore, for large grains, the FE solution can be expected to behave more like the Reuss case, since the boundary value problem tends towards a series connection of grains with respect to the tensile direction.

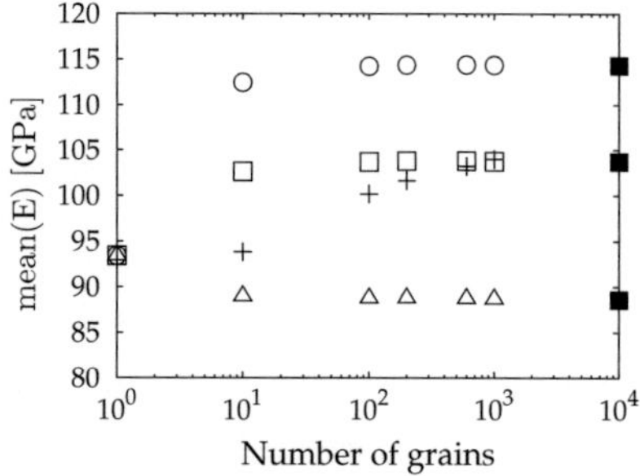

Figure 5.8: Mean value of Young's modulus of Stabilor®G for different numbers of grains in the sample: ○ arithmetic mean, △ harmonic mean, □ singular approximation with geometric mean as reference material, ■ corresponding isotropic estimates, + FE simulations

For gold, Young's modulus is known to be bounded by the extremal values located at $\langle 100 \rangle$ and $\langle 111 \rangle$ crystal direction, which is reconfirmed by Figure 5.9. Both simultaneously form a rough estimate for the largest possible range of dispersion for each of the numbers of grains in the specimen.

The finite element calculations, described in the previous section, show a decreasing scatter band with an increasing number of grains, which is represented by the extremal values appearing in the simulations in Figure 5.7. The magnitude of the scattering of Young's modulus predicted numerically agrees with experimental findings. However, the simulations predict a slightly smaller

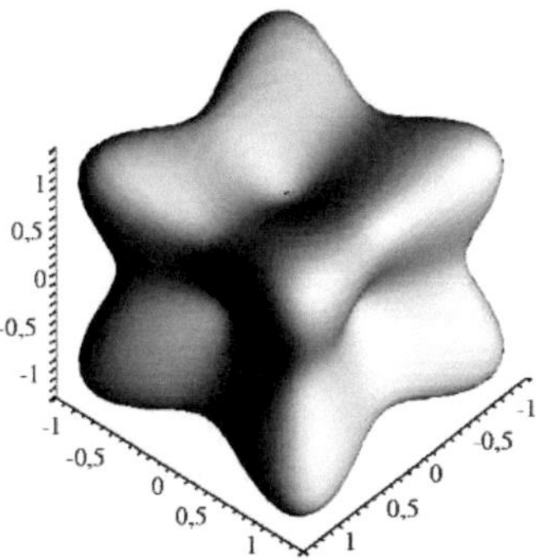

Figure 5.9: Graphical representation of Young's modulus $E(n)$ of Stabilor®G

standard deviation ($\sqrt{\mathrm{var}(E^{\mathrm{FE}})} = 1.2$ GPa) compared to the one determined by experiments ($\sqrt{\mathrm{var}(E^{\exp})} = 2$ GPa) for 600 grains in the specimen.

Both the simple bounds as well as the singular approximation depend on the fourth-order texture coefficient. The norm of $\mathbb{V}'$ for uniform discrete orientation distributions as a function of the number of orientations is shown in Figure 5.10. It can be seen that for uniform orientation distributions, the norm tends to zero for large grain numbers. Figure 5.10 also confirms the result that for 600 grains the mean value of Young's modulus does not differ much from the isotropic estimate since the norm of the texture coefficient is already quite small.

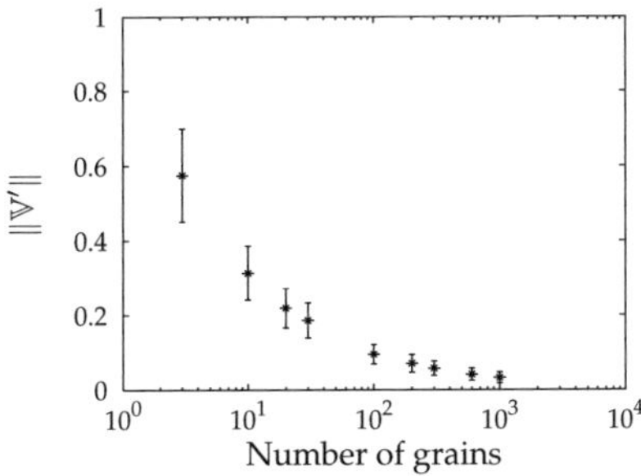

Figure 5.10: Norm of the texture coefficient subject to the number of grains in the sample

For 600 grains in the specimen, the singular approximation predicts a standard deviation of $\sqrt{\mathrm{var}(E^{\mathrm{SA}})} = 1.1$ GPa independent of the chosen reference material. Hence, the accuracy of both methods is identical, although the numerical effort is much larger for finite element simulations. The one-point bounds do not reproduce the mean value of Young's modulus. This result has been expected, since the material shows a significant amount of single crystalline anisotropy which necessarily implies that these bounds fail. In contrast to the mean value prediction it is, however, remarkable, that the one-point bounds give good results for the prediction of the scattering of Young's modulus over the full range of grain numbers. The value for 600 grains in the specimen determined using one-point bounds is $\sqrt{\mathrm{var}(E^{\mathrm{V,R}})} = 1.0$ GPa.

Generalization to materials with cubic crystal symmetry

In order to understand the behavior of Young's modulus in microspecimens more generally, its mean value and standard deviation are examined for three more cubic materials (gold, aluminum and nickel). Similar finite element simulations as for Stabilor®G are performed for these three materials, the single crystalline elastic constants of which are given in Table 5.2. Furthermore, the dimensionless Zener ratio $Z = \lambda_3/\lambda_2$ is given as well, which can be used to quantify the degree of anisotropy. For $Z = 1$, the single crystal is isotropic, otherwise, it is anisotropic. It can be seen that aluminum shows the smallest and gold the largest degree of anisotropy among the examined materials. Due to the proportional scaling of the material parameters of gold, the Zener ratios of gold and Stabilor®G are identical.

	gold	nickel	aluminum
C_{1111}[GPa]	185	251	107
C_{1122}[GPa]	158	150	61
C_{1212}[GPa]	39.7	124	28
Z	2.9	2.5	1.2

Table 5.2: Elastic constants of the examined materials (Simmons and Wang, 1971; Paufler and Schulze, 1978)

Figure 5.11(a) shows the mean value of Young's modulus with respect to the number of grains. In a double logarithmic plot, the behavior of Young's

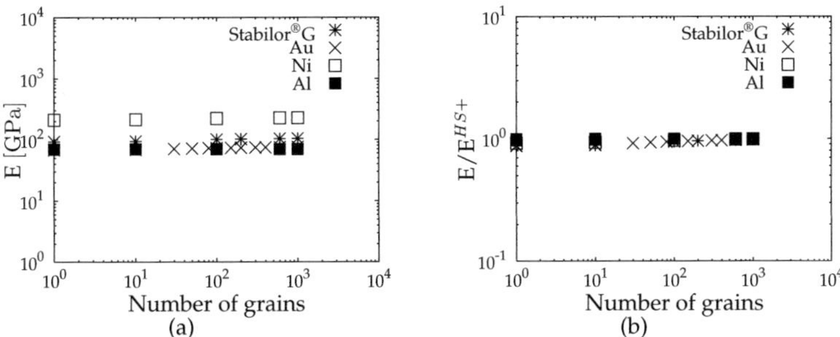

Figure 5.11: (a) Young's modulus and (b) normalized Young's modulus as a function of the number of grains in the sample for different fcc materials

modulus with respect to the number of grains is characterized by a straight line with a slight slope for all of the examined materials. This slope can be explained by the so called first-order size effect (Geers et al., 2006).

Using the upper Hashin-Shtrikman bound, to achieve a dimensionless representation of the results, yields a generalization of the curves (Figure 5.11(b)). The choice of E^{HS+} is reasonable, since it has been shown that this bound is approached by increasing the grain number.

In Figure 5.12(a), the results of the finite element analysis concerning the scattering of Young's modulus are shown. In a double logarithmic plot, for all the examined materials, the scattering turns out to decrease linearly with an increasing grain number. Such a decreasing behavior of the scattering of the elastic constants has already been described in Kanit et al. (2003) for a composite of two linear elastic phases, wherein it is shown that this linear decrease holds for different types of boundary conditions.

Additionally, it can be seen here that the slope of the curves for different materials with cubic crystal symmetry is very similar (Figure 5.12(a)). Based on this finding, the question arises whether a master curve exists being valid for all cubic materials with weak or moderate anisotropy.

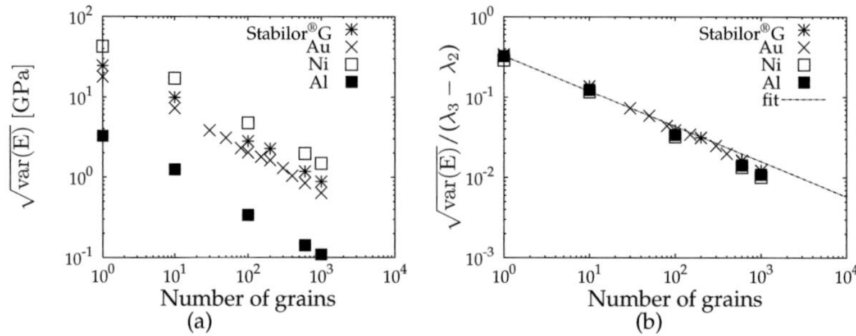

Figure 5.12: (a) Standard deviation and (b) normalized standard deviation of Young's modulus subject to the number of grains in the sample for different fcc materials

If the standard deviation is normalized by the amount of anisotropy $\lambda_3 - \lambda_2 = \lambda_2(Z - 1)$, such master curve is obtained, which captures quite accurately the scattering of Young's modulus for all the materials (see Figure 5.12(b)). Hence, the scattering of Young's modulus as a function of the grain number N can be approximated for cubic materials by

$$\frac{\sqrt{\mathrm{var}(E)}}{\lambda_3 - \lambda_2} \approx k_1 N^{-k_2}, \qquad k_1 = 0.326, \qquad k_2 = 0.438, \tag{5.100}$$

wherein the constants k_1 and k_2 have been identified by least squares fitting.

Analogously, using the singular approximation, the predicted fitting parameters are $k_1 = 0.321$ and $k_2 = 0.470$, which implies a slightly smaller value for the standard deviation of Young's modulus for large grain numbers compared to predictions from FE simulations. When using the self-consistent approach (SC), the results are pretty similar to the ones obtained by SA. Fitting the curve of decreasing normalized standard deviation yields $k_1 = 0.333$ and $k_2 = 0.478$. The fitted curves for the three approaches are shown in Fig. 5.13. For the purpose of quantifying the deviation, exemplarily for 10^4 grains of Nickel, the SA and SC would consistently predict $E = 225 \pm 0.6\,\mathrm{GPa}$ and the FE approach $E = 225 \pm 0.8\,\mathrm{GPa}$.

For cubic crystal aggregates, the results imply that the statistics of the apparent properties can be determined based on anisotropic effective medium approximations which take into account the statistics of the microstructure.

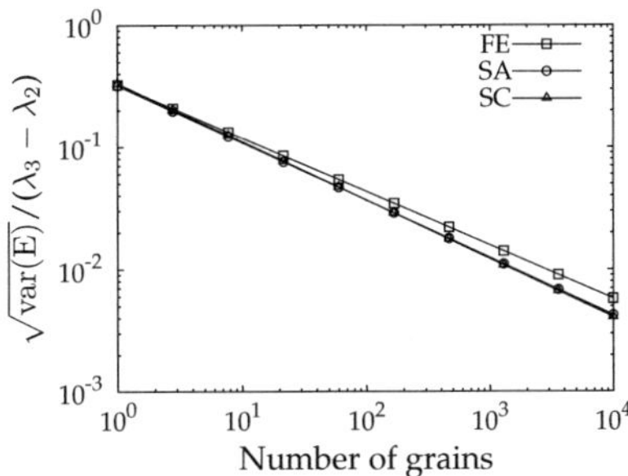

Figure 5.13: Numerically (FE) vs. semi-analytically (SA and SC) predicted normalized standard deviation of Young's modulus with respect to the number of grains in the sample

Remarkably, this is true for the whole range of grain numbers, from the single crystal to the polycrystalline aggregate, even though the assumptions of these approaches are not met for volume elements smaller than the RVE.

5.7.2 Anisotropy due to crystallographic and morphological texture

The elastic anisotropy induced by the manufacturing process is usually signif-icant. In contrast to the investigations in the preceding section, in this study the number of grains is large enough to be statistically representative. The macroscopic anisotropy, therefore, only results from the texture in the specimen. This section aims to discuss separately the influence of crystallographic and morphological texture and their combination. Partly, this section summarizes the results published in Jöchen et al. (2010) and Jöchen and Böhlke (2010).

Effective elasticity of a RVE with a single elliptic inclusion

This paragraph aims to quantify the influence of the shape of the inclusion on the effective elastic response of a RVE. For this study, one single elliptical inclusion in a matrix is investigated, both, inclusion and matrix, possess-ing an isotropic material behavior. Inspired by metal matrix composites (MMC), the elastic constants of the particle and the matrix are taken to be $E_M = 70\mathrm{GPa}, \nu_M = 0.32$ and $E_I = 375\mathrm{GPa}, \nu_I = 0.22$, which represents an Al_2O_3 particle in an aluminum matrix. Such a material has a strong phase contrast in

Young's moduli, being $1 : 5$. Three different particle geometries are examined,

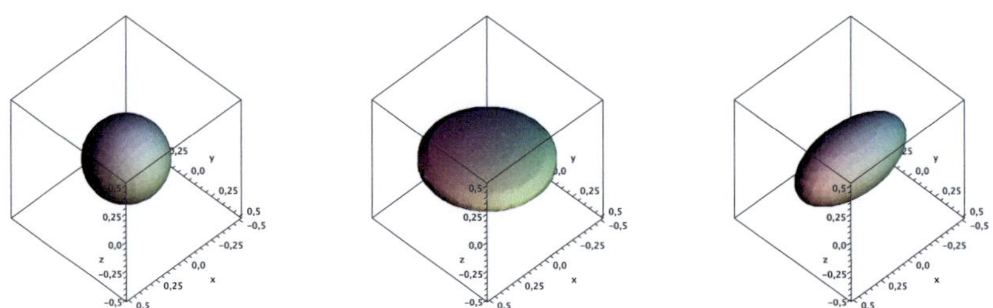

Figure 5.14: Three elliptical inclusion shapes: sphere, disc, needle

all being ellipsoids, or more specifically spheroids. In case of the sphere, all three half-axes are equal, for the disc it is assumed that its shape is a resultant of compressing a sphere in z-direction, whereas the needle is a sphere being elongated in x-direction (see Fig. 5.14). The volume fraction of the particle is $V/V_{RVE} = 0.1$ in all three cases, resulting in an aspect ratio of 2 in case of the needle and 4 in case of the disc.

For the evaluation of the effective properties of such a RVE, in a first step, computational homogenization based on finite elements is applied. Therefore, the RVE cube is meshed in a structured manner using 125000 linear hexahedral elements. Similar to the procedure for generating a Voronoi microstructure on a structured grid in section 5.7.1, in each integration point it is determined whether the associated volume belongs to the matrix or inclusion phase. Once more, this results in multi-phase elements as discussed before. Then, the six orthogonal load cases are applied (cf. section 5.6) using periodic boundary conditions to compute the effective stiffness $\bar{\mathbb{C}}^{FE}$ of the RVE. Note that in this case, due to the assumed cubic periodicity of the inclusion position, the effective material behavior becomes cubic regardless of the isotropic material behavior of the constituents and the spherical shape.

In a second step, as a representative for the semi-analytical schemes, the self-consistent estimate is applied being able to account for ellipsoidal morphology.

The SC estimate and the FE computations are compared to the Voigt and Reuss bounds, which, however, are insensitive to any morphological details.

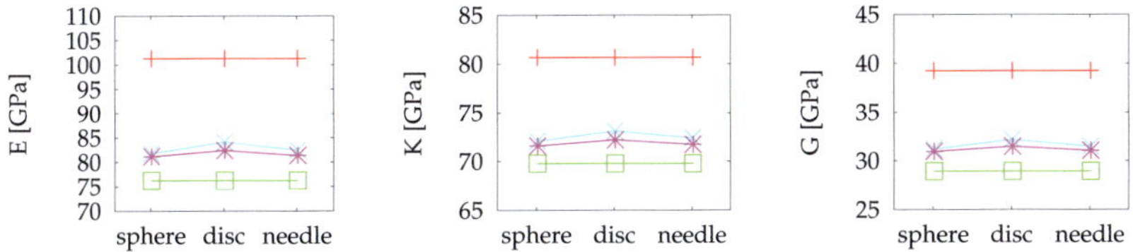

Figure 5.15: Influence of inclusion shape on elastic parameters: Young's modulus $\bar{E}$, bulk modulus $\bar{K}$, shear modulus $\bar{G}$; red: Voigt, green: Reuss, magenta: SC, cyan: FEM

At first, the isotropic projections of the obtained effective stiffness tensors $\bar{\mathbb{C}} = \bar{\mathbb{C}}^V, \bar{\mathbb{C}}^R, \bar{\mathbb{C}}^{SC}, \bar{\mathbb{C}}^{FE}$ are computed giving the effective isotropic estimates of the bulk modulus $3\bar{K} = \bar{\mathbb{C}} \cdot \mathbb{P}_1^I$ and the shear modulus $10\bar{G} = \bar{\mathbb{C}} \cdot \mathbb{P}_2^I$. The associated Young's modulus is than given by $\bar{E} = 9\bar{K}\bar{G}/(3\bar{K} + \bar{G})$. The results of these evaluations are depicted in Fig. 5.15. As expected due to the character of an isotropic projection of an anisotropic stiffness tensor, the influence of anisotropic inclusion morphologies is not adequately represented by examining these elastic parameters. However, it becomes obvious that the one-point bounds predict an unsatisfying broad range of admissible effective behavior due to the large material contrast of the phases. On the other hand, when examining the entries of the effective stiffness tensor (Fig. 5.16), the strong influence of the shape of the inclusion on the effective elastic behavior becomes apparent. While the sphere leads to an overall isotropic material behavior, the disc and needle induce macroscopic anisotropies only due to their shape, since the material behavior of the phases is isotropic. The variation of the elastic constants due to the shape is up to 10% in this example. Compared to the full-field simulations, the SC method yields very good results within only a fraction of the time. The presented results confirm the necessity to account for phase morphologies in addition to volume fraction distributions and the constitutive behavior of the single phases, to reliably predict the overall material behavior of heterogeneous materials.

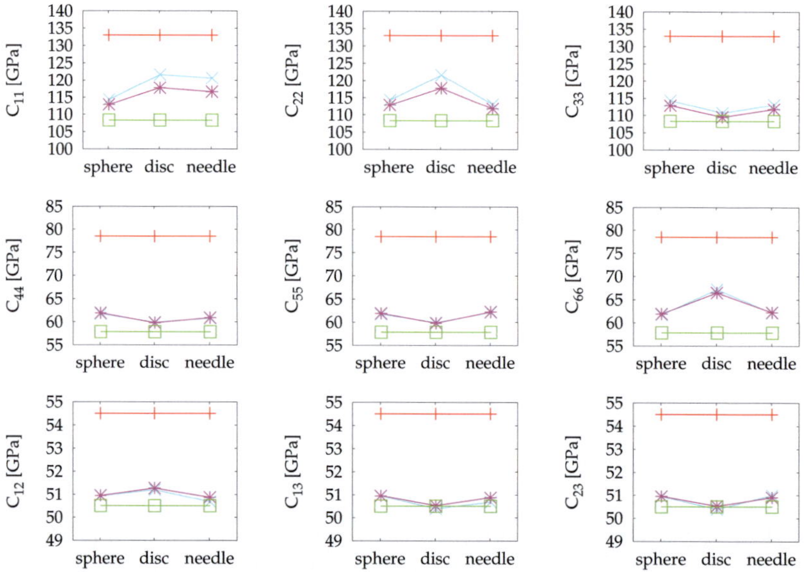

Figure 5.16: Influence of inclusion shape on effective elastic constants of the elasticity tensor in modified Voigt notation; red: Voigt, green: Reuss, magenta: SC, cyan: FEM

Effective elastic behavior of rolled sheets

The preceding investigations emphasized the need for the incorporation of morphological aspects into the predictions of the effective behavior of heterogeneous materials. This paragraph documents investigations of the elastic behavior of rolled sheet metals consisting of cubic single crystals. In terms of the influence of crystallographic orientation distribution, among others, this problem has already been investigated by Bunge (1968b). During the rolling process, however, the grains of the polycrystalline aggregate undergo severe deformations leading not only to significant lattice reorientations (crystallographic texture) but also to tremendous grain shape changes (morphological texture). In contrast to section 7.1, where the rolling process with the associated inelastic behavior of the sheet is simulated, this paragraph investigates the elastic loading of a rolled sheet. The crystallite orientation distribution entering this elastic analysis has been generated using a uniform orientation distribution and a rigid-viscoplastic Taylor-type simulation (Taylor (1938); assumption of homogeneous deformation throughout the whole crystal aggregate; see also section 6.1.1) to produce the CODF of the rolled sheet (see also Böhlke et al. (2009)). In these rolling simulations, the prescribed deformation gradient is given by

$$\bar{F} = \lambda e_R \otimes e_R + \lambda^{-1} e_N \otimes e_N + e_T \otimes e_T, \tag{5.101}$$

with the rolling, normal and transverse direction of the sheet being denoted with e_R, e_N and e_T, respectively. For 1000 uniformly distributed crystals, the discrete (100) pole figures are depicted in Fig. 5.17 for different degrees of thickness reduction. The shape, which is consistent with deforming a sphere by the

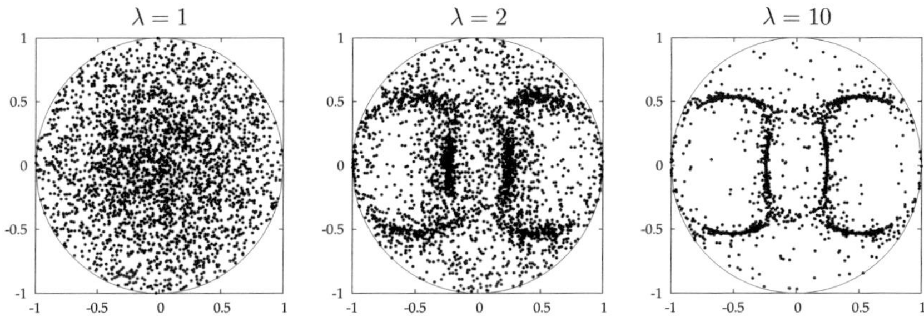

Figure 5.17: (100) pole figures for different degrees of thickness reduction by rolling: 0%, 50% and 90% (from left to right); (taken from Jöchen et al., 2010)

deformation gradient from Eq. (5.101), will be used in the respective mean field approaches to represent the mean grain shape.

Exemplarily, a sheet metal made of copper with the single crystalline constants $C_{1111} = 168\,\text{GPa}$, $C_{1122} = 121\,\text{GPa}$ and $C_{1212} = 75\,\text{GPa}$ has been examined. In a first step, a sheet consisting of 1000 grains that has been rolled to 90% of thickness reduction is investigated. To show the only influence of the crystallographic texture, Young's modulus in the sheet metal plane is evaluated by using the one-point and HS bounds as well as the self-consistent estimate, assuming a spherical grain shape (Fig. 5.18(a)). The crystallographic texture has a strong influence on the elastic behavior in the sheet metal plane, resulting

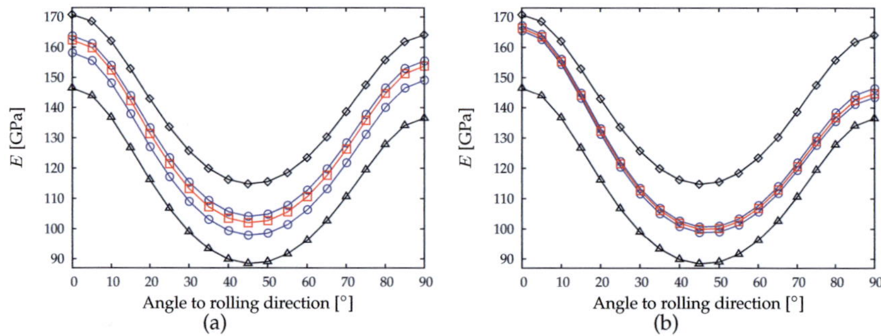

Figure 5.18: Young's modulus in sheet metal plane: (a) isotropic two-point statistics, (b) anisotropic two-point statistics; $\diamond$ Voigt ; $\triangle$ Reuss ; $\circ$ HS$^{\pm}$; $\square$ SC

in a variation of Young's modulus of about 50 GPa. When additionally taking into account morphological texture (Fig. 5.18(b)), the HS and SC schemes show that the anisotropy of Young's modulus becomes even more pronounced yielding to an additional inclination of the whole curve. Furthermore, the HS bounds are much tighter, when the morphology is included. To illustrate

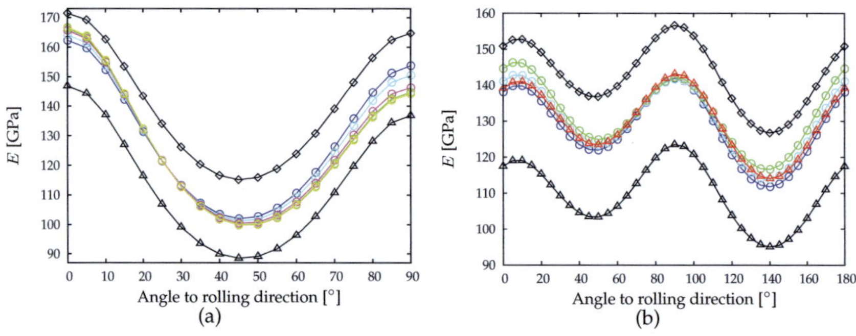

Figure 5.19: Young's modulus in sheet metal plane: (a) 90% thickness reduction of 1000 grains (b) 50% thickness reduction of 50 grains: $\diamond$ V ; $\triangle$ R ; $\circ$ SC $\alpha = 1$; $\circ$ SC $\alpha = 2$; $\circ$ SC $\alpha = 5$; $\circ$ SC $\alpha = 10$; $\circ$ SC $\alpha = 20$; $\triangle$ FE

the potential of the morphology incorporation even more, the self-consistent method is used. When taking the morphology tensor entering $\mathbb{P}_0$ (Eq. (5.51)) as $\boldsymbol{A}^{-1} = \alpha e_R \otimes e_R + \alpha^{-1} e_N \otimes e_N + e_T \otimes e_T \sim \bar{\boldsymbol{F}}$ according to the form of the macroscopic deformation, the factor α has been varied to show its influence (Fig. 5.19(a)). Note that the variation of α, surely, is unphysical and artificial, since the crystallographic texture is kept constant. For the considered range $\alpha \in [1, 20]$, the largest change in Young's modulus is observable in transverse direction being approximately 10 GPa.

Additionally, to show the validity of the applied mean field approaches, a finite element computation has been carried out. Therefore, a Voronoi type mesh with an inhomogeneous metric has been produced (Böhlke et al., 2009; Jöchen and Böhlke, 2009c; Fritzen et al., 2009) to model a rolled grain structure. For this comparative computation, 50 grains out of the 1000 have been assigned to the Voronoi microstructure. In this case, the volume fraction distribution associated with the FE model was also used in the self-consistent method. Fig. 5.19(b) depicts the results of this study. It becomes obvious that due to the small grain number, the macroscopic material behavior is not yet orthotropic, since a symmetry in the Young's modulus distribution with respect to the transverse direction (90°) is not perfectly achieved. Furthermore, the finite element result is in between the SC estimate for $\alpha = 1$ and $\alpha = 2$. This is reasonable, since the compressed Voronoi grains do not exhibit the perfect ellipsoidal shape of the SC approach. Nevertheless, this study confirms the validity of the assumption of ellipsoidal grain shape according to the macroscopic deformation in the SC method.

It can be concluded that the mean field theories offer a great potential for computing the overall behavior of crystal aggregates possessing crystallographic and morphological texture. While the simple bounds represent the boundaries for the real material behavior, they, however, are far apart (especially in this example due to the large elastic anisotropy of copper) and can not account for morphology aspects of the microstructure at all. It has been shown, that the self-consistent estimate lies within the Hashin-Shtrikman bounds for both, isotropic and anisotropic two-point statistics. The crystallographic texture has a major influence on the overall elastic behavior, but nevertheless, the morphological texture is not negligible. Kröner (1977) already explained mathematically with the help of the distance of bounds, why changes in the lattice orientations lead to larger variations in the effective behavior than changes in grain shape. However, it is worth incorporating the morphological information, since especially for large deformations (cf. the 90% thickness reduction case) it can completely change the characteristic behavior in terms of inclining the curves. It should be noted that the example with an aggregate of 50 grains being not statistically homogeneous shows that the SC estimate (and, therefore, also the HS bounds) are applicable also for such oligocrystalline aggregates since the results show to be consistent with the finite element results. This finding agrees with the one of the preceding microcomponent study.

Chapter 6

Physically Non-linear Homogenization Methods

6.1 Bounds

The motivation for studying bounds has already been discussed for the class of linear material behavior and can, analogously, be transmitted to non-linear problems. For bounding the non-linear material behavior, usually the concepts proven successful in linear elasticity are extended. In general, however, only the simple approaches lead to both, upper and lower bounds. The higher-order theories mainly yield one-sided bounds, as is also briefly summarized in the following subsections.

6.1.1 Taylor and Sachs bounds

Similar to the Voigt approach in linear elasticity, Taylor (1938) assumes uniform strain throughout the polycrystal. This model satisfies kinematic compatibility and provides an upper bound to the effective potential, which has been proven by Bishop and Hill (1951). On the other hand they showed that when the static compatibility is trivially fulfilled assuming overall uniform stress in the heterogeneous material (e.g., Hosford and Galdos, 1990), resulting in a violation of the kinematic compatibility, a lower bound is obtained. The latter method is ofter referred to as the Sachs model (Sachs, 1928), although Sachs originally assumes proportional (not uniform) loading which, in general, satisfies neither strain compatibility nor stress equilibrium (Leffers, 1979; Kocks et al., 1998; Gilormini and Bréchet, 1999). Nevertheless, according to the habitual use, the assumption of homogeneous stress is also entitled Sachs-type behavior throughout this work.

Since the assumptions of homogeneous stress or strain are artificial and rarely exist in reality, several modifications are suggested, which do in general not generate bounds but estimates. Many models investigate in weakening the Taylor model, being the 'Relaxed constraint theories' (e.g., Kocks and Chandra, 1982; Van Houtte, 1982), where some shear components are allowed to deviate from the average values[15] or the LAMEL, the ALAMEL (Van Houtte et al., 1999, 2005) and the GIA models (Crumbach et al., 2001), where not every single grain but a stack of grains is subjected to the macroscopic deformation. Analogously, modifications of the Sachs model (Leffers, 1979; Pedersen and Leffers, 1987) have been introduced, in which additional random stresses are assigned to each grain so as to induce multislip. This is contrarily to the original model, which inherently induces single slip.

6.1.2 Talbot-Willis- and Ponte Castañeda-type bounds

Following the work of Willis (1983), who generalized the Hashin-Shtrikman principle to non-linear problems, Talbot and Willis (1985) derived a one-sided bound based on a homogeneous linear comparison medium and Talbot and Willis (1994a,b, 1995) also developed the other bound, which necessitates the introduction of non-linearity to the homogeneous comparison material. The classical Taylor and Sachs bounds are included in the Talbot-Willis procedure for the extremal cases of the comparison material behaving infinitely stiff or compliant. The Talbot-Willis bounds are two-point bounds.

Ponte Castañeda (1992), as well as deBotton and Ponte Castañeda (1995) in the context of polycrystals, derived a class of bounds, which are based on linear heterogeneous comparison materials, also named comparison composites. This procedure contains the Talbot-Willis bounds as a special case, but is more general as it offers the potential to derive also higher-order bounds as, e.g., Beran-type bounds (Beran, 1965). The latter are not available by applying the Talbot-Willis principle. Ponte Castañeda's procedure has the same shortcoming as the first Talbot-Willis-approach as it delivers only one-sided bounds (see also Ponte Castañeda and Suquet, 1998). Talbot and Willis (1997) overcame this shortcoming by using a non-linear behavior of the comparison material to ob-

---

[15]Note that the RC models are usually not very general since the components to be relaxed depend on the load case. A generalized RC model is introduced by Van Houtte and Rabet (1997) preventing this shortcoming, but actually belonging to the class of self-consistent models (Van Houtte et al., 1999).

tain the other side bound, whereas also three-point information is incorporated in their approach.

6.2 Incremental self-consistent scheme

When considering the elasticity of polycrystalline materials, the self-consistent model (cf. section 5.5) shows to predict very realistic results. Due to this fact it is obvious trying to extent the self-consistent approach to the class of non-linear material behavior. Kröner (1961) and Budiansky and Wu (1962) were the first to suggest an incremental linearization of the material response so as to apply the SC scheme in each time step. Furthermore, Hill (1965a,b)[16] proposed a pretty similar approach except for a softer grain interaction. Both schemes are compared by Hutchinson (1970), who judged Hill's method more realistic. The difference in both approaches is the relation between the stress and strain fluctuations in the heterogeneous material. Departing from the phase average of the strain in phase α

$$\varepsilon_\alpha = \bar{\varepsilon} - \mathbb{P}_0[\boldsymbol{p}_\alpha - \langle \boldsymbol{p} \rangle], \qquad \boldsymbol{p}_\alpha = \boldsymbol{\sigma}_\alpha - \mathbb{C}_0[\varepsilon_\alpha], \tag{6.1}$$

where the stress in this phase is a non-linear function of the strain $\boldsymbol{\sigma}_\alpha = \boldsymbol{\sigma}_\alpha(\varepsilon_\alpha)$, the following fluctuation relation can be derived

$$\boldsymbol{\sigma}_\alpha - \bar{\boldsymbol{\sigma}} = \mathbb{L}[\varepsilon_\alpha - \bar{\varepsilon}], \qquad \mathbb{L} - \mathbb{C}_0 - \mathbb{P}_0^{-1}. \tag{6.2}$$

$\mathbb{L}$ is called Hill's constrained tensor (Hill, 1965a) and in the self-consistent sense $\mathbb{C}_0$ has to be chosen to describe the effective medium. The difference in the Kröner-Budiansky-Wu model, compared to Hill's approach, lies in the computation of $\mathbb{L}$. While the former assumes an equal interaction in elasticity and non-linearity so that the constraint tensor is computed based on the effective elastic modulus $\mathbb{L}^{KBW} = \mathbb{L}(\mathbb{C}_0 = \bar{\mathbb{C}}_{elastic})$, in Hill's approach, each single phase is embedded in a matrix possessing the overall instantaneous modulus leading to $\mathbb{L}^{Hill} = \mathbb{L}(\mathbb{C}_0 = \bar{\mathbb{C}}_{inst})$. Since during plastic deformation the instantaneous (or continuum modulus) is smaller compared to the elastic stiffness tensor, Hill's method yields a softer interaction between the phases. Due to the state dependence of the continuum moduli of the phases, however, $\mathbb{L}$ is not constant

---

[16]The extension of Hill's approach to rate-dependent materials has been discussed by Nemat-Nasser and Obata (1986).

in Hills model, the opposite is valid for the Kröner-Budiansky-Wu approach which, therefore, renders the method computationally more efficient.

The behavior predicted by the incremental SC scheme, nevertheless, shows to be stiffer than the real material response simply explainable by the incremental linearization. Therefore, Jöchen and Böhlke (2009a,b) proposed to compute $\mathbb{L}$ on the basis of the algorithmic consistent moduli of the phases (see, e.g., Simó and Hughes (1998)), which are even softer than the respective continuum moduli (see, e.g., Doghri and Ouaar (2003)) to end up with a slightly softer response than the one obtained by Hill's method. This choice is obvious due to the time discretization of the deformation process implying already a certain linearization.

6.3 Non-linear Hashin-Shtrikman-type homogenization

6.3.1 Generalization of the linear HS scheme

The incremental self-consistent approach shows to yield too stiff results when using either the continuum or consistent tangent operators (e.g., Lebensohn and Tomé, 1993; González and LLorca, 2000). This is possibly due to the assumption of piecewise homogeneous fields, which is not applicable in case of inelastic deformations. The latter generally induce strong field heterogeneities inside the phases. Some better results using the incremental self-consistent scheme are obtained when using isotropization of the tangent operators during the evaluation procedure (Doghri and Ouaar, 2003; Jiang and Shao, 2009). To soften the responses compared to all these aforementioned self-consistent approaches while maintaining the assumption of piecewise constant polarizations in the volume V_α of a phase or a grain α

$$p(x) = \sum_{\alpha=1}^{N} \chi_\alpha(x) p_\alpha, \qquad \chi_\alpha(x) = \begin{cases} 1 & \forall x \in V_\alpha \\ 0 & \text{otherwise} \end{cases}, \tag{6.3}$$

in this work, the linear Hashin-Shtrikman procedure is generalized in terms of allowing a non-linear stress response in phase (or grain) α being $\sigma_\alpha = \sigma_\alpha(\varepsilon_\alpha)$. Furthermore, in opposition to the procedure optimizing $\mathbb{C}_0$ to obtain bounds, in the proposed method, $\mathbb{C}_0$ is perceived as being a free parameter imposing a certain statistics on the stress and strain fields.

The polarization tensor $\mathbb{P}_0$ (see also Eq. (5.51) and the section 5.2.3)

$$\mathbb{P}_0 = \frac{1}{4\pi\det(\boldsymbol{A})}\iint\limits_{\partial \mathcal{S}} \mathbb{H}(\mathbb{C},\boldsymbol{n})(\boldsymbol{n}\cdot(\boldsymbol{A}^{-\mathsf{T}}\boldsymbol{A}^{-1}\boldsymbol{n}))^{-3/2}\,\mathrm{d}S(\boldsymbol{n}) \qquad (6.4)$$

is used to relate the stress polarization and strain fluctuation similar as in the incremental self-consistent scheme (see, e.g., Eq. (6.1)). The interaction law in the present case is given by

$$\boldsymbol{\sigma}_\alpha - \bar{\boldsymbol{\sigma}} = \mathbb{L}[\boldsymbol{\varepsilon}_\alpha - \bar{\boldsymbol{\varepsilon}}], \qquad \mathbb{L} = \mathbb{C}_0 - \mathbb{P}_0^{-1}, \qquad (6.5)$$

which is identical to Eq. (6.2), with the difference of a non-committed comparison material $\mathbb{C}_0$.

For cubic crystals, the spherical part of the stiffness tensor is isotropic so that the interaction law in Eq. (6.5) reduces to an equation relating the deviatoric parts of the stress and strain tensors. In the case of an isotropic comparison medium, Eq. (6.5) only depends on a scalar factor[17]

$$\boldsymbol{\sigma}'_\alpha - \bar{\boldsymbol{\sigma}}' = -\frac{\lambda_2^0(3\lambda_1^0 + 4\lambda_2^0)}{2(\lambda_1^0 + 3\lambda_2^0)}(\boldsymbol{\varepsilon}'_\alpha - \bar{\boldsymbol{\varepsilon}}'). \qquad (6.6)$$

In this equation, it can easily be noticed that the proposed scheme inherently includes the Taylor method of homogeneous strain by choosing $\lambda_2^0 \to \infty$ and the Sachs method of homogeneous stress by $\lambda_2^0 \to 0$. This fact is also true for Eq. (6.5) when choosing the comparison medium as infinitely stiff or compliant, respectively.

6.3.2 Localization rule for the spin

Only in case of validity of the assumption of an isotropic phase shape (spherical morphology of the phases), the spin is homogeneously distributed in the heterogeneous material, so that the macroscopic spin is identically transferred to each of the N single phases, i.e., $\boldsymbol{\omega}_\alpha = \bar{\boldsymbol{\omega}}$ ($\alpha = 1,\ldots,N$). Contrarily, if the shape of the phases is anisotropic, then the stress polarization distribution influences the spin distribution. This suggests using a localization rule incorporating the

---

[17]Note that also Berveiller and Zaoui (1979) use a scalar factor for a similar fluctuation relation between stresses and plastic strains being $-\lambda_2^0(3\lambda_1^0 + 4\lambda_2^0)/(5(\lambda_1^0 + 2\lambda_2^0))$, where the comparison medium is chosen in the self-consistent sense with tangent or secant moduli.

skew part of the second derivative of Greens function according to Lipinski and Berveiller (1989); Adams and Field (1991); Lebensohn and Tomé (1993); Berveiller (2001), which obviously can be derived by using the skew part of the gradient of Eq. (5.24). In this sense, the fluctuation of the infinitesimal spin is given by

$$\boldsymbol{\omega}_\alpha = \bar{\boldsymbol{\omega}} - \mathbb{P}_0^A[\boldsymbol{p}_\alpha - \langle \boldsymbol{p} \rangle], \tag{6.7}$$

where the microstructural tensor $\mathbb{P}_0^A$ is defined similar to $\mathbb{P}_0$ by

$$\mathbb{P}_0^A(\mathbb{C}_0, \boldsymbol{A}) = \frac{1}{4\pi \det(\boldsymbol{A})} \iint\limits_{\partial \mathcal{S}} \mathbb{H}^A(\mathbb{C}_0, \boldsymbol{n})(\boldsymbol{n} \cdot \boldsymbol{A}^{-\top}\boldsymbol{A}^{-1}\boldsymbol{n})^{-3/2} \, \mathrm{d}S(\boldsymbol{n}), \tag{6.8}$$

with $\mathbb{H}^A = \mathbb{I}^A(\boldsymbol{N}\square(\boldsymbol{n} \otimes \boldsymbol{n}))\mathbb{I}^S$ including the identity on skew-symmetric second-order tensors $\mathbb{I}^A$.

6.3.3 Aspects of numerical implementation

The first question arising when trying to implement Eq. (6.5) for a computation involving large deformations as, e.g., a rolling process, is the definition of the transition of the kinematic variables so as to use the small strain formulation of section 6.3. Therefore, assuming that due to the time discretization of the deformation process, the deformation gradient at the beginning $\boldsymbol{F}_n$ and at the end of the time step $\boldsymbol{F}_{n+1}$ are given, the deformation during the time step is $\Delta \boldsymbol{F} = \boldsymbol{F}_{n+1}\boldsymbol{F}_n^{-1}$. One possibility to determine the kinematic variables in small strain context is to use straightforwardly the increment of the displacement gradient and its symmetric and skew-symmetric parts

$$\Delta \boldsymbol{H} = \Delta \boldsymbol{F} - \boldsymbol{I}, \quad \Delta \boldsymbol{\varepsilon} = (\Delta \boldsymbol{F} + \Delta \boldsymbol{F}^\top)/2 - \boldsymbol{I}, \quad \Delta \boldsymbol{\omega} = (\Delta \boldsymbol{F} - \Delta \boldsymbol{F}^\top)/2. \tag{6.9}$$

In this case, the principle of invariance under superimposed rigid-body motions (PISM, see, e.g., Bertram (2005)) is not fulfilled. This can easily be proved by setting $\Delta \boldsymbol{F} \in SO(3)$ in Eq. (6.9). Also for very small rotations in each of the small time steps one has to be aware that, nevertheless, strains (and resulting stresses) can accumulate for rigid-body rotations reaching non-negligible values at the end of pure large rotations. Another approach overcoming this shortcoming is the approximation of $\Delta \boldsymbol{\varepsilon} \approx \Delta t \boldsymbol{D} = \Delta t \, \mathrm{sym}(\boldsymbol{L})$ with the velocity gradient $\boldsymbol{L} = \dot{\boldsymbol{F}}\boldsymbol{F}^{-1}$. When using the fact that $\boldsymbol{D}$ can be given by

$$\boldsymbol{D} = \mathrm{sym}(\boldsymbol{L}) = \frac{1}{2}\boldsymbol{F}^{-\top}\dot{\boldsymbol{C}}\boldsymbol{F}^{-1}, \tag{6.10}$$

with the time derivative of the right Cauchy-Green tensor $C = F^\mathsf{T}F$ and discretizing $\dot{C} = (C_{n+1} - C_n)/\Delta t$, then

$$\Delta\varepsilon = \frac{1}{2}\left(I - \Delta F^{-\mathsf{T}}\Delta F^{-1}\right).$$
(6.11)

Obviously, using this definition, rigid-body rotations do not induce any strains even for large rotations, so that Eq. (6.11) is an incrementally objective approximation (see also Simó and Hughes (1998)). In this case, one has to be aware of pure stretches which could induce spin since

$$\Delta\omega = \Delta t W = \Delta t(L - D) = \frac{1}{2}\left(I - 2\Delta F^{-1} + \Delta F^{-\mathsf{T}}\Delta F^{-1}\right).$$
(6.12)

When not taking into account the discretization used in the approximation of D, then no spin is induced at all for pure stretches since

$$\Delta\omega = \Delta t W = \Delta t\,\mathrm{skw}(L) = \frac{1}{2}\left(\Delta F^{-\mathsf{T}} - \Delta F^{-1}\right).$$
(6.13)

The approximation (6.11), therefore, is exact in case of pure rotations and (6.13) in case of pure stretches.

The second aspect to discuss is the type of solution of Eq. (6.5). Two different approaches have been implemented within the scope of this thesis. On the one hand, aiming to solve the problem as accurately as possible, an implicit scheme has been used. In this case, Eq. (6.5) has been rewritten

$$G_\alpha = \mathbb{L}^{-1}(\sigma_\alpha - \bar{\sigma}) - \varepsilon_\alpha + \bar{\varepsilon} = 0,$$
(6.14)

with $\alpha = 1,\ldots,N$. Solving Eq. (6.14) with the Newton method, the tensorial entries of the required Jacobian read

$$\mathbb{J}_{\alpha\beta} = \frac{\partial G_\alpha}{\partial \varepsilon_\beta} = (\delta_{\alpha\beta} - c_\beta)\mathbb{L}^{-1}\mathbb{C}_\beta^{alg}(\varepsilon_\beta) - \delta_{\alpha\beta}\mathbb{I}^S.$$
(6.15)

Note that $\delta_{\alpha\beta} = 1 \; \forall \alpha = \beta$ and $\delta_{\alpha\beta} = 0$ otherwise. The dimension of the entire Jacobian, therefore, is, in general, $6N \times 6N$, with N the number of phases. Due to the constraint $\mathrm{d}\bar{\varepsilon} = \sum_\alpha c_\alpha \,\mathrm{d}\varepsilon_\alpha$, the phase average of strain for one phase can be eliminated. Furthermore, when the phases of the polycrystal possess cubic crystal symmetry, Eq. (6.14) can be formulated in the sense of the strain deviators, so that the dimension of the Jacobian is at least $5(N - 1) \times 5(N - 1)$. As a consequence of the general necessity of small time steps for this scheme and the large dimensions of the Newton equation in case of a representative number of grains/phases, the above presented procedure results in large computational times.

On the other hand, Eq. (6.5) has been solved for the strains per phase being given explicitly, however, as a function of the inelastic strains. In this approach, $\sigma_\alpha = \mathbb{C}_\alpha[\varepsilon_\alpha - \varepsilon_\alpha^{in}]$ with the inelastic part of the strain tensor ε_α^{in} is used, so that Eq. (6.5) can be alternatively written as

$$\mathbb{M}_\alpha^{-1}[\varepsilon_\alpha] - \langle \mathbb{M}^{-1}[\varepsilon] \rangle = \mathbb{P}_0 \mathbb{C}_\alpha[\varepsilon_\alpha^{in}] - \mathbb{P}_0 \langle \mathbb{C}[\varepsilon^{in}] \rangle. \tag{6.16}$$

Note that $\mathbb{M} = (\mathbb{I}^S + \mathbb{P}_0 \delta \mathbb{C})^{-1}$, as already used in section 5.4. It can be easily verified that the ansatz for the strain in phase α

$$\varepsilon_\alpha = \mathbb{M}_\alpha \langle \mathbb{M} \rangle^{-1}[\bar{\varepsilon}] + \mathbb{M}_\alpha \mathbb{P}_0 \mathbb{C}_\alpha[\varepsilon_\alpha^{in}] - \mathbb{M}_\alpha \langle \mathbb{M} \rangle^{-1} \langle \mathbb{M} \mathbb{P}_0 \mathbb{C}[\varepsilon^{in}] \rangle \tag{6.17}$$

fulfills Eq. (6.16). The first addend in Eq. (6.17) represents exactly the elastic localization of the strain, whereas the last two terms give the influence of the inelastic deformation. Contrary to the above mentioned Newton scheme with Eq. (6.15), in this approach, the strains are given explicitly and do not require the solution of a coupled system of equations. Nevertheless, iterative steps can be performed for equilibrating the total and inelastic strains in order to improve the solution.

The third important numerical aspect is directed towards the implementation of the proposed generalized scheme at the integration points of finite elements for two-scale simulations. In this case, it is necessary to compute the algorithmic tangent of the method for numerical efficiency. As discussed before, the usage of the explicit scheme is much more efficient so that only this method can prospectively be applied in two-scale problems. If the scheme is utilized in its strictly explicit sense, then the inelastic part of the strains are a function of the previous time step $\varepsilon_\alpha^{in} = \varepsilon_\alpha^{in}(t_n)$, so that the algorithmic tangent is exactly given as the effective medium in linear elasticity by

$$\bar{\mathbb{C}}_{alg} = \frac{\partial \bar{\sigma}}{\partial \bar{\varepsilon}} = \langle \mathbb{C} \mathbb{M} \rangle \langle \mathbb{M} \rangle^{-1}. \tag{6.18}$$

If the scheme is used with $\varepsilon_\alpha^{in} = \varepsilon_\alpha^{in}(t_{n+1})$ with iteratively computing the equilibrium of ε_α and ε_α^{in}, then the algorithmic tangent reads

$$\bar{\mathbb{C}}_{alg} = \frac{\partial \bar{\sigma}}{\partial \bar{\varepsilon}} = \langle \mathbb{C}_{alg} \mathbb{M} \rangle \langle \mathbb{M} \rangle^{-1}. \tag{6.19}$$

6.4 Applications within the geometrically linear framework

This section is only devoted to physically non-linear small strain applications as an extension of the topics addressed in section 5.7. Homogenization in the context of finite deformations, especially the examination of predicting texture evolution, is extensively but separately discussed in section 7.

6.4.1 Anisotropy of oligocrystals

Extending the investigations of the linear response of microspecimens made of Stabilor®G (section 5.7.1) to predict the non-linear behavior, especially the initial yielding is examined. Therefore, an initial critical resolved shear stress $\tau_{C0} = 190\,\mathrm{MPa}$ is assumed for all slip systems, the value being determined on the basis of the initial yield stress of the macroscopic stress strain curves (Fig. 5.4(c)) and the well known Taylor value 3.06 (see, e.g., Hutchinson, 1976). Hardening is neglected due to the experimental indication, and the strain rate sensitivity is assumed to be $m = 20$. The R_{p02} value, being the tensile stress at 0.2% plastic strain, has been identified experimentally by five tensile tests to be $R_{p02} = 621 \pm 41\mathrm{MPa}$. Numerically, larger sets of 200 test specimens with randomly generated microstructure for each discrete number of grains have been examined. Fig. 6.1 shows the results generated by the finite element method. Figure 6.1 (left) gives the results for R_{p02}. As expected, the scattering of the yield stress decreases with increasing grain number. It becomes obvious for 600 grains that the mean value is captured very well, whereas the scattering is underestimated compared to the experimental results. More experimental results would be required to analyze and identify the reason for the large experimental scattering, and probably even some information on the hardening behavior of different slip systems.

Additionally, this figure gives analytical values for the single crystal and the polycrystal with infinite number of grains. Since these data hold for initial yielding and 0.2% plastic strain represents a state lying already comparatively far in the plastic region, a sharper criterion being the technical elastic limit at 0.01% plastic strain is depicted on the right in Fig. 6.1. Obviously, by using R_{p001}, it is confirmed that the numerical results tend between the Taylor and Sachs bounds for an increasing grain number. The same sampling as for the finite element method has been investigated by the incremental self-consistent scheme and

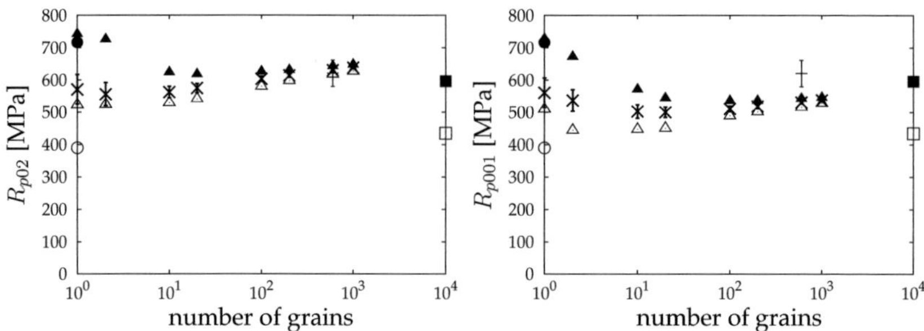

Figure 6.1: (left) yield stress R_{p02}, (right) technical elastic limit R_{p001}; symbols denote: $\times$ average R_{px}, ■ isotropic Taylor upper bound for initial yielding, □ isotropic Sachs lower bound for initial yielding, ▲ maximum R_{px}, △ minimum R_{px}, ○ minimal single crystal initial yield stress, ● maximal single crystal initial yield stress, + experimental results for R_{p02} charge Z2 (Auhorn (2005))

the Taylor approach (Fig. 6.2). Both schemes result in stiffer material response compared to the full field simulations, nevertheless, the incremental SC scheme is superior to the Taylor method. The effect of computational efficiency of the mean-field theories compared to the full-field approach becomes even more pronounced than for the examinations in section 5.7.1, when physically non-linear material behavior (in the present case crystal plasticity) is involved in the computations. For example, the full-field simulations with 600 grains required a factor of 200 in computational time of the incremental self-consistent scheme, the Taylor method inherently being even more efficient. Since the results of the application of the mean-field theories are comparable to the results of the FE approach, the former are clearly recommended for statistically investigating the macroscopic behavior of micro-heterogeneous materials.

6.4.2 Anisotropy due to crystallographic and morphological texture

As an extension of the investigations of the elastic behavior of rolled copper sheets in section 5.7.2, the latter are now analyzed in the inelastic regime (also see Böhlke et al. (2009) building the basis of this study). Therefore, tensile specimens of the sheets that have been rolled to 50% thickness reduction are virtually cut in rolling and in transverse direction. These specimens consisting of 50 grains are elongated by 1% to examine the stress-strain response. The

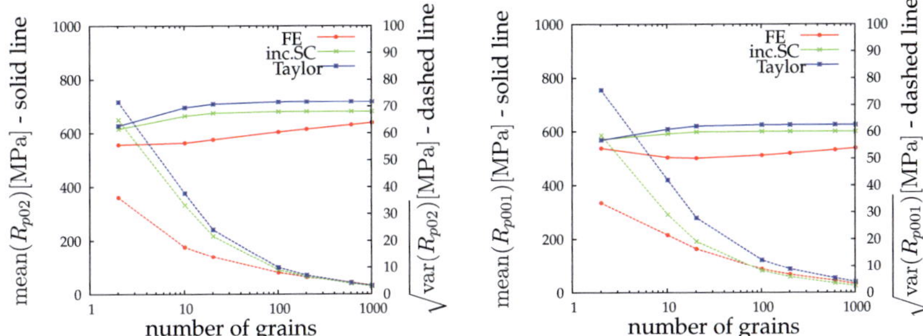

Figure 6.2: Numerical vs. analytically predicted (left) yield stress R_{p02}, and (right) technical elastic limit R_{p001}

tensile test is carried out in a volume-preserving isotropic manner, meaning that the two normal strains in the cross section are prescribed to be half the negative tensile strain. The critical resolved shear stress is assumed to be constant $\tau_{C0} = 42$ MPa, and the rate-independency is approximated by setting $m = 80$. Figures 6.3 and 6.4 show the stress-strain curves for the tensile test in

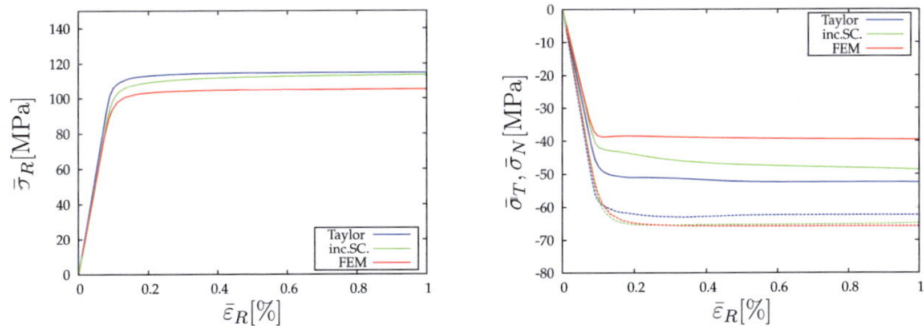

Figure 6.3: Stress-strain curves of a tensile test in rolling direction for a rolled specimen of 50 grains and 50% thickness reduction

rolling and transverse direction, respectively, where the finite element method, the Taylor scheme as well as the incremental self-consistent approach have been applied. Fig. 6.5 shows the FE models for the anisotropy investigations which

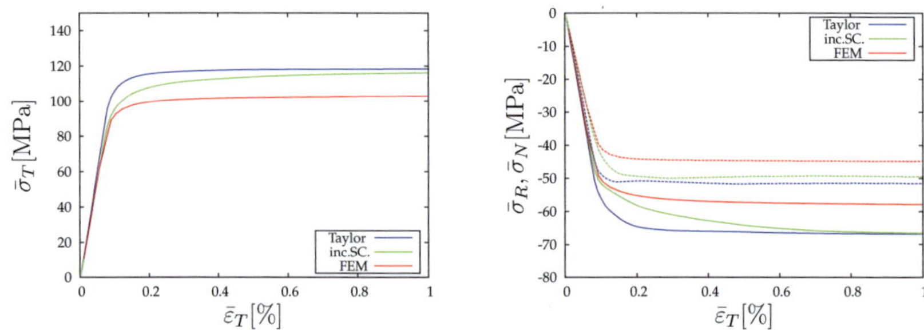

Figure 6.4: Stress-strain curves of a tensile test in transverse direction for a rolled specimen of 50 grains and 50% thickness reduction

Figure 6.5: FE-models of the specimen consisting of 50 grains: raw (left, referred to as $M = 0$) and rolled to 50% thickness reduction (right, referred to as $M = 1$)

have already been used in Böhlke et al. (2009) with the meshing technique of Fritzen et al. (2009).

In all cases, the stiffest response in tensile direction is obtained by the Taylor approach and the softest response by the FEM. The gap between both is larger for the tensile test in transverse direction (Fig. 6.4 (left)). The incremental SC gives results in between the Taylor and FEM prediction, approaching the Taylor result for increasing elongations. This fact is known and has already been described in the corresponding theoretical section 6.2. For the response in the directions perpendicular to the tensile direction, the behavior is more anisotropic in case of the tensile test in rolling direction. Among all the methods,

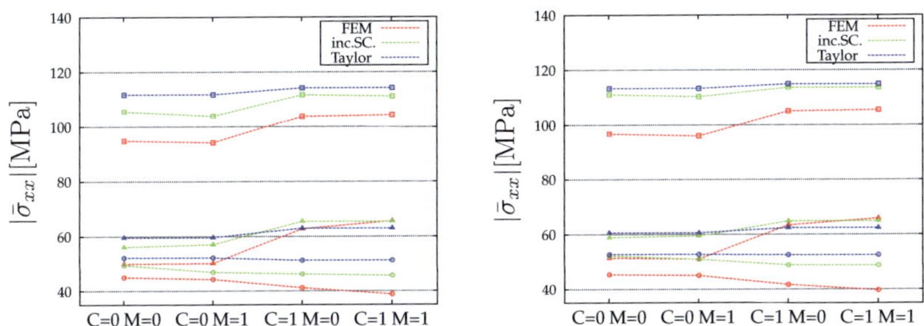

Figure 6.6: Influence of crystallographic (C) and morphological (M) texture on effective stress at (left) 0.3% elongation and (right) 1% elongation in the tensile test in rolling direction: $\square\ \sigma_{11}$, $\circ\ \sigma_{22}$, $\triangle\ \sigma_{33}$

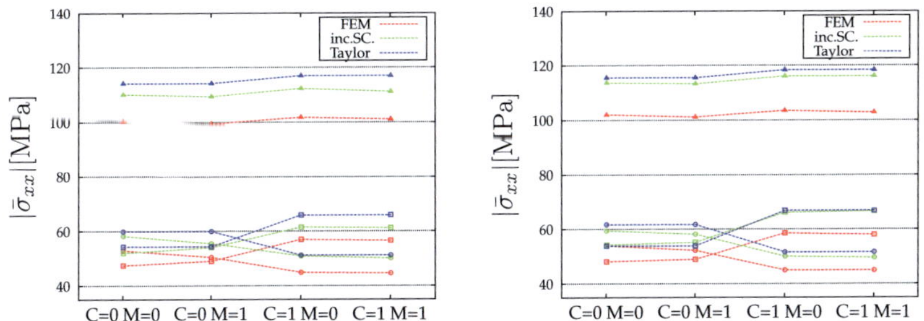

Figure 6.7: Influence of crystallographic (C) and morphological (M) texture on effective stress at (left) 0.3% elongation and (right) 1% elongation in the tensile test in transverse direction: $\square\ \sigma_{11}$, $\circ\ \sigma_{22}$, $\triangle\ \sigma_{33}$

the Taylor scheme predicts the most isotropic behavior in the cross section of the tensile specimen, while the FEM gives the most anisotropic response. In case of the tensile test in transverse direction, the prediction of cross sectional anisotropy of the stress response is almost the same for all three methods.

To separate the influence of crystallographic and morphological texture on the stress response during the deviatoric tensile tests, the four different configurations have been tested by switching the crystallographic (C) and morphological (M) texture on (1) and off (0). The results of these investigations are depicted in Figs. 6.6 and 6.7 for 0.3% and 1% of elongation in rolling and transverse direction, respectively. Once more it becomes obvious that the incremental self-consistent result approaches the Taylor response for larger deformations, however, the latter cannot reflect morphological aspects. In general, the influence of crystallographic texture is much more pronounced than the influence of morphological texture. Compared to the full-field simulations, the incremental self-consistent method predicts the analogous behavior under crystallographic as well as morphological changes. For the tensile test in rolling direction (Fig. 6.6), the crystallographic texture causes an increase in the tensile stress, while the morphological texture causes a decrease. For the rolling texture ($C = 1$), the incorporation of anisotropic grain morphology results in an increasing cross sectional anisotropy which is more severe when using FEM than the incremental scheme. Especially for small strains (Fig. 6.6 (left)), the prediction of the influence of crystallographic texture is improved by the incremental scheme compared to the Taylor result when taking the FEM as reference. In case of the tensile test in transverse direction (Fig. 6.7), the effect of the crystallographic texture is less pronounced compared to the tensile test in rolling direction. In this tensile test, the crystallographic texture causes the ratio of cross sectional stresses to revert. The influence of the morphology is much more pronounced for $C = 0$ than for $C = 1$, especially for the small strain diagram (Fig. 6.7 (left)). In all these cases, the incremental scheme gives the same trend of texture influence as the FEM and is significantly favored compared to the Taylor result, especially in the elastic-plastic transition.

Chapter 7

Geometrically Non-linear Applications of the Non-linear Hashin-Shtrikman-type Scheme

7.1 Texture prediction

In this section, the non-linear Hashin-Shtrikman type homogenization method is applied in the large deformation regime to predict texture evolution. A raw data set for aluminum AA3104 is, therefore, taken from Delannay et al. (2002) and has been representatively reduced to a small number of 285 (and 926 for comparison) texture components using the method presented in section 4.1. The reduced data sets are applied in the sequel to predict texture due to rolling, elongation and shear. Although experimental data of a deformation texture is only available for the rolling process, the other typical deformation modes are also applied in order to evaluate the potential of the proposed model. The single crystalline material parameters of AA3104 used for the simulations in this subsection are given in Table 7.1.

λ_1	λ_2	λ_3	τ_{C0}	$\tau_{C\infty}$	Θ_0	$\dot{\gamma}_0$	τ_D	m
229 GPa	46 GPa	56 GPa	40 MPa	60 MPa	1000 MPa	$10^{-3}\,\mathrm{s}^{-1}$	13 MPa	20

Table 7.1: Material parameters for aluminum

7.1.1 Rolling texture

Similar to the investigations of Delannay et al. (2002) (notice also the erratum Delannay et al., 2003), the considered process is cold rolling up to 55% thickness reduction described by the velocity gradient of plane strain compression

$$L = \frac{\dot{\gamma}_0}{\sqrt{2}} \begin{pmatrix} 1 & 0 & 0 \\ 0 & 0 & 0 \\ 0 & 0 & -1 \end{pmatrix}_{\{e_i \otimes e_j\}} . \tag{7.1}$$

The experimental textures before and after rolling measured by X-ray diffraction are shown in Fig. 7.1. Three specific sections of the CODF at $\varphi_2 = 45°$, $65°$, $90°$

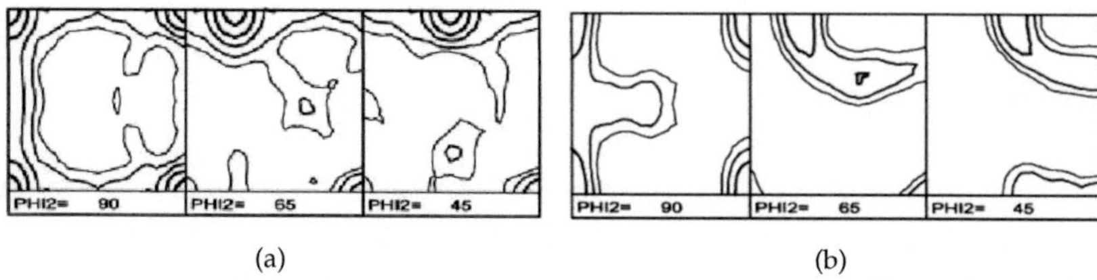

(a)   (b)

Figure 7.1: Experimental textures of AA3104: (a) 0% th.red. and (b) 55% th.red., contours at $1/2/4/8/16$ (Delannay et al., 2002)

are depicted, since they include typical texture components as copper (Cu), brass (B), S (S), Goss (G) and cube (C) (see Fig. 7.2). In the sequel, the non-linear

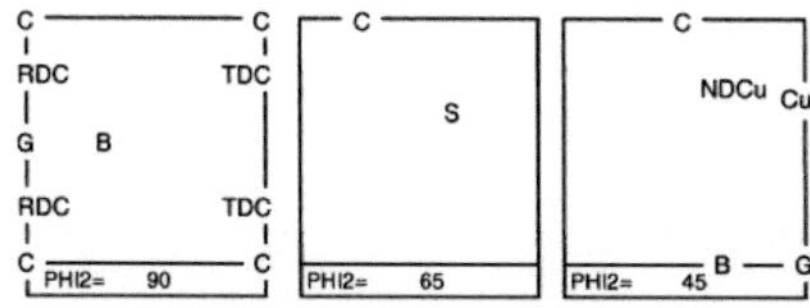

Figure 7.2: Location of typical texture components in three suitable CODF sections (Delannay et al., 2002)

Hashin-Shtrikman-type scheme will be used to predict the CODF of the rolled AA3104 specimen. A detailed discussion on the influence of different aspects of the method on the CODF prediction, especially with regard to typical textural fibers, follows in the next paragraphs.

Influence of choice of $\mathbb{C}_0$, morphological texture and spin localization

For the evaluation of the rolling texture prediction with the non-linear Hashin-Shtrikman type model, three pronounced textural fibers are evaluated being the α-, β- and θ-fibers (e.g., Hirsch and Lücke, 1988b; Delannay et al., 1999). While the α- and θ-fibers are fixed in Euler space giving the CODF intensities along $\varphi_1 \in [0°, 90°]$, $\Phi = 45°$, $\varphi_2 = 90°$ and $\varphi_1 = 0°$, $\Phi \in [0°, 90°]$, $\varphi_2 = 90°$, respectively, the β-fiber describes the location and intensity distribution of the CODF maxima in CODF sections at $\varphi_2 \in [45°, 90°]$. The latter is also referred to as 'skeleton line' (Bunge, 1993). Note that in this work, the β-fiber is actually interpreted as connecting the maxima of the CODF, which does not necessarily mean that it includes the brass, S and copper components in either case. The latter interpretation of the β-fiber is sometimes used in other studies. Experimental

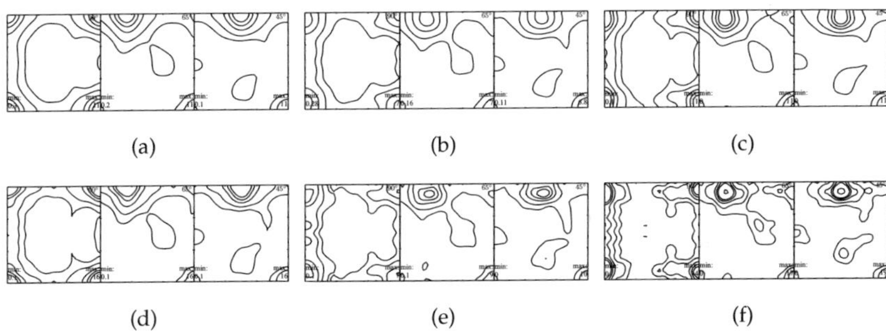

(a)   (b)   (c)

(d)   (e)   (f)

Figure 7.3: Approximated CODF at 0% th.red: (a, d) 10000 discrete orientations, (b, e) 926 texture components and (c, f) 285 texture components; upper row 7° and lower row 5° half-width; contours at 1/2/4/6/8/16

data of the CODF before and after rolling has been provided in a format containing the CODF values on a grid with 5° steps. Furthermore, the initial texture of AA3104 was also available in a discretized version containing 10000 discrete Euler angle triples with identical weights. Using MTEX[18] to draw a continuous CODF representation out of the discretized data, the results of Fig. 7.3(a) for a half-width of 7° (the same half-width used in Delannay et al.

---

[18]with 'De-la-Vallee-Poussin' kernel

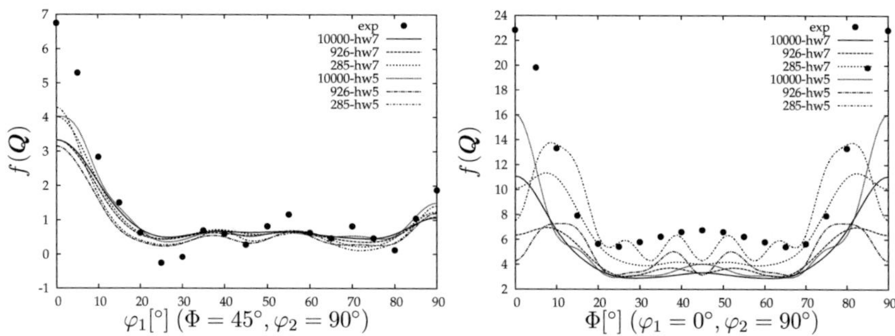

Figure 7.4: α- (left) and θ-fiber (right) of approximated CODFs of the initial texture

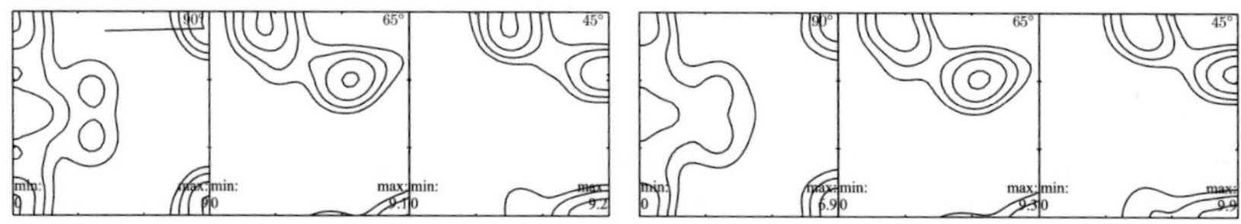

Figure 7.5: CODF slices for 55% th.red., a factor of $k = 0.06$ and the AMS method: 285 (left) and 926 texture components (right)

(2002)) are obtained. Although the approximated CODF visually complies with the measurement (cf. Fig. 7.1(a)), the extremal values differ. Decreasing the half-width to $5°$ increases the extremal value only to some extent (Figs. 7.3(d) and 7.4). Since the discretized data set has been reduced to a smaller number of texture components for subsequent simulations, one needs to be aware of the fact, that capturing the exact values of the experimentally determined CODF is nearly impossible. This has already become obvious by the aforementioned comparisons between experimental and discretized textures with the entire 10000 components. In the sequel, two different reduced data sets containing 285 and 926 texture components of the initial texture are used in the rolling simulations. The CODF approximations of the initial texture by these two data sets are depicted in Fig. 7.3 as well. Additionally, the α- and θ-fiber of the experimental CODF and the discretizations with 10000 grains as well as 926 and 285 texture components for the two chosen half-widths are shown in Fig. 7.4. Besides the already mentioned lack of capturing the exact maximum values, the qualitative slopes of the fibers are well reproduced by all three discretizations.

Decreasing the half-width leads to an increasing waviness of the CODF for the low-dimensional approximations. Furthermore, it becomes obvious that due to the data reduction technique using averaging in boxes of the orientation space (cf. Chapter 4), the strong cube component being located at the corners of the orientation space (as well as at both ends of the θ-fiber) is being shifted towards the interior by some degrees. This is an inherent deficiency of the reduction method which, understandably, has a stronger impact for smaller numbers of boxes and, thereby, larger averaging domains.

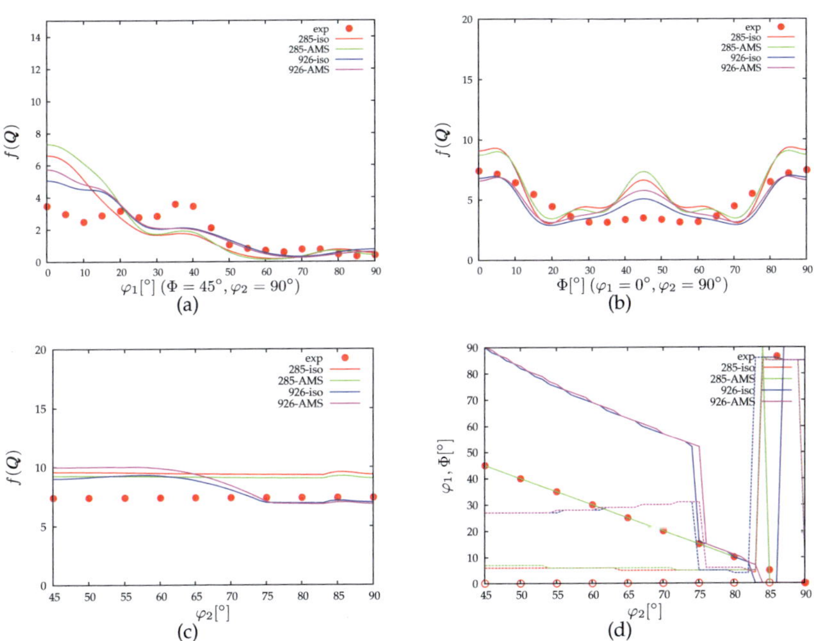

Figure 7.6: (a) α-, (b) θ-fiber, (c) intensity and (d) location of β-fiber of approximated CODFs of the AA3104 texture at 55% th.red. for a factor $k = 0.06$ and a half-width of $7°$; in (d) solid lines and symbols denote the φ_1-location and dashed lines and void symbols the Φ-location

The texture evolution has been computed using the non-linear Hashin-

Shtrikman scheme with on the one hand, an isotropic comparison medium (iso) and on the other hand, an anisotropic comparison medium (A) with additional morphology information in the ellipsoidal sense (M) and the appropriate spin localization (S). In this paragraph, all of the latter three aspects are either accounted for (AMS) or not (iso), while in the next subsections, the single influences of A, M and S are discussed separately. In case of iso, the isotropic comparison medium was chosen to be the arithmetic mean of the elastic stiffnesses of the grains (isotropic Voigt average) multiplied by a factor $k \in [0, \infty]$, in order to vary between Sachs- and Taylor-type behavior $\mathbb{C}_0 = k \sum_\alpha c_\alpha \mathbb{C}_\alpha^{\text{iso,elast}}$. In the case A, the arithmetic mean of the algorithmic stiffnesses has been used as the anisotropic basis of the comparison medium $\mathbb{C}_0 = k \sum_\alpha c_\alpha \mathbb{C}_\alpha^{\text{algo}}$. The entire results for the fiber predictions for the two reduced data sets (285 and 926 texture components, respectively) using three different half-widths, different grades k between the Sachs- and Taylor-type comparison material, as well as the methods iso and AMS, are given in the appendix A.1. In general, the Goss component is overestimated by very small and very large values k. Furthermore, for $k \rightarrow 0$ and $k \rightarrow \infty$, the brass component is under- and overestimated, respectively. The exact cube component intensity is too weak, which on the one hand is due to the meta-stability resulting of the high symmetry of this component (Delannay et al., 2002), and on the other hand, can be an artifact of the aforementioned reduction technique, which already shifts the extremal CODF intensity at cube position of the initial texture to a small extent. Note that this aspect of the experimental rolling texture could also generally not be captured satisfactorily by the homogenization methods used in Delannay et al. (2002). Even the comparatively highly resolved FEM prediction of the rolling texture showed a stronger copper than cube behavior. In the present investigations, it is peculiar that the correct location of the β-fiber from experiments can only be completely predicted correctly for the small number of texture components (285) with values of $k > 0.05$. For smaller k, the location of the β-fiber agrees only for $\varphi_2 > 65°$, which then holds also for 926 texture components.

The computations with 926 texture components nearly universally resulted in a very typical β-fiber including Cu and S contrarily to the experiments, but for $\varphi_2 \rightarrow 90°$, a stronger cube (or slightly rotated cube) than brass is predicted. Besides the non-matching maximum location, the latter behavior agrees with the experimental findings (see, e.g., Fig. 7.1).

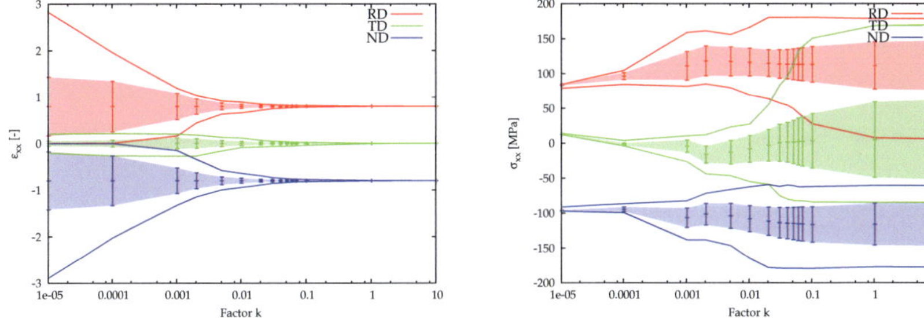

Figure 7.7: Distribution of normal stress and pseudo-strain components with respect to the factor k; lines denote extremal values, shaded zone denotes range of mean value and standard deviation

It has been found that the best agreement between experiment and simulation is found for using either 285 or 926 texture components, a factor of $k = 0.06$ and the AMS method (Figs. 7.5 and 7.6). Note, that the β-fiber location is, however, predicted differently for the two different numbers of texture components, by giving the experimental location for 285 components, and the typical B-S-Cu run for 926 components. Although the fibers are generally reproduced very well, the copper and S components are, however, overestimated compared to the experiments.

The distributions of phase averages of normal stresses and pseudo-strains[19] with respect to the factor k are exemplarily depicted in Fig. 7.7 for using 926 texture components and the AMS method. Since the characteristics are similar for the iso case or the smaller number of texture components, their distributions are not shown separately. The lines denote the extremal values of the stress and pseudo-strain components in the aggregate, while the mean values and standard deviations are given by errorbars and are additionally visualized by the shaded zone. It becomes obvious that a Taylor-type behavior with homogeneous strain is obtained approximately for a factor $k \geq 1$, while a factor of $k \leq 10^{-5}$ tends towards the homogeneous stress state (Sachs-type behavior). However, perfect stress homogeneity is not reached for $k = 10^{-5}$. Smaller values of k are not tested in this study since in that case, very small

---

[19]Note that in this geometrically non-linear context, the ε is actually not a strain measure, since it can not be computed from the total displacements but is the accumulation of strain increments each belonging to the updated configuration due to the deformation.

time steps would be required for solving the boundary value problem and, therefore, lead to huge computational costs. The representation of the stress and pseudo-strain distributions by mean values and standard deviations is only appropriate if a Gaussian distribution is present. Fig. 7.8 shows the histograms

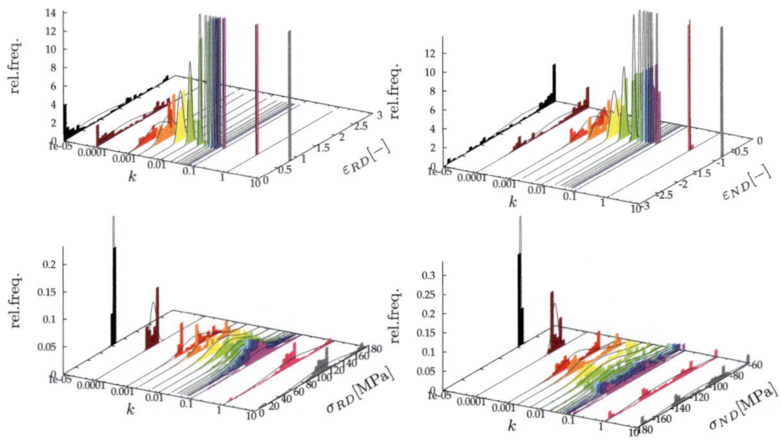

Figure 7.8: Pseudo-strain and stress distributions in RD and ND with respect to k

of pseudo-strains and stresses in RD and ND for 926 texture components and the AMS method additionally depicting the Gaussian bell-shaped curves (for more histograms, see appendix A.2). While σ_{RD} approximately behaves in a Gaussian manner with clusters around the mean value, the pseudo-strain distribution in RD and ND for small k and the stress distribution of σ_{ND} for larger k rarely show normal distributions. In general, it is thus necessary to examine the complete histograms and not only mean values and standard deviations to characterize the types of distribution.

To judge the computational efficiency of the non-linear Hashin-Shtrikman scheme with respect to the application of the Taylor method, Tab. 7.2 specifies the computational time needed for the rolling simulation normalized by the time needed for the Taylor homogenization on the same computer. Note that for different factors k, a different number of increments was used to solve the

k	10^0	10^{-1}	10^{-2}	10^{-3}
increments	1	2	2	10
Time: iso	1.19	2.35	2.65	11.29
Time: A	3.06	6.45	7.10	55.65
Time: AM	3.55	6.77	7.84	46.31
Time: AMS	6.61	13.22	21.61	86.45

Table 7.2: Comparison of computational time with respect to Taylor scheme

problem, which is indicated by the second row of the table. The number of increments is also normalized with respect to the number of time increments required by the Taylor simulation. As expected, the iso method represents the most efficient procedure, due to the explicit representation of the microstructural tensor. In this case, if normalized by the number of increments, the solution can be obtained in nearly the same time as the Taylor method ($t_{\text{HS,iso}}^{\text{CPU}} \approx 1.3\, t_{\text{Taylor}}^{\text{CPU}}$). Among all methods, AMS is computationally most expensive due to the need for the numerical computation of a second microstructural tensor being responsible for controlling the spin localization.

Comparison with finite element computation

In order to additionally compare the results of the non-linear homogenization scheme with own finite element results and, therefore, to eliminate the influence of slightly different material laws or parameters applied in own computations and those of Delannay et al. (2002), a fully resolved artificial microstructure using the Voronoi-type algorithm of section 3.3.2 is created for the case with 285 grains. Only one realization of this microstructure has been generated and examined here, knowing full well that different realizations having varying proximity relations could also result in slight differences when examining the CODFs. The FE model of the RVE with periodic microstructure before and after rolling is shown in Fig. 7.9. The macroscopic deformation is prescribed using periodic boundary conditions on the model consisting of 64000 hexahedral elements with linear ansatz functions.

The CODF obtained in the rolling simulation by evaluating the data of all 512000 Gauss points is depicted in Fig. 7.10. It can be seen that the predicted CODF

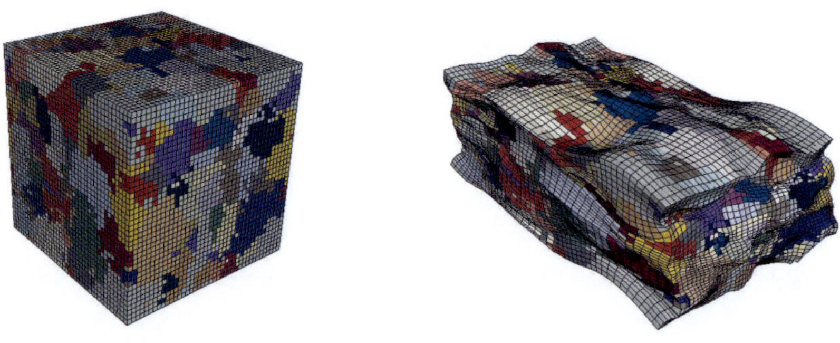

Figure 7.9: RVE with 285 grains at 0% and 55% th. red. during rolling

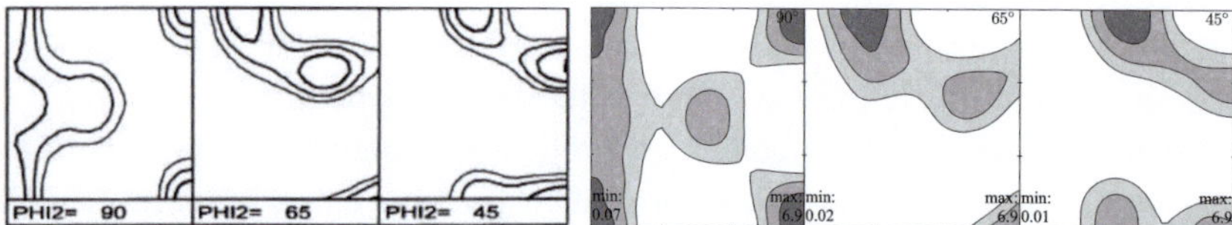

Figure 7.10: CODF at 55% th.red. obtained by finite element simulations: (left) results of Delannay et al. (2002); (right) own results

mainly complies with the one computed by Delannay et al. (2002), especially for the strong C component. In contrast, the Cu component is much less pronounced in the own FEM results, but this is in better agreement with the experimental results of Delannay et al. (2002) (cf. Fig. 7.1(b)).

Furthermore, the distribution of (pseudo-)strains and stresses are evaluated in Fig. 7.11. When comparing the FEM results with the distributions predicted by the non-linear Hashin-Shtrikman scheme it becomes obvious that for smaller k, the (pseudo-)strain distributions are consistently predicted and for larger k, the stress distributions agree well. The best agreement in the stress distribution is predicted for $k = 0.06$, which complies with the preceding findings for the agreement of the textural fibers in the experiment and texture simulation (cf. Fig. 7.6). However, the pseudo-strain distribution in this case is too homogeneous compared to FEM results.

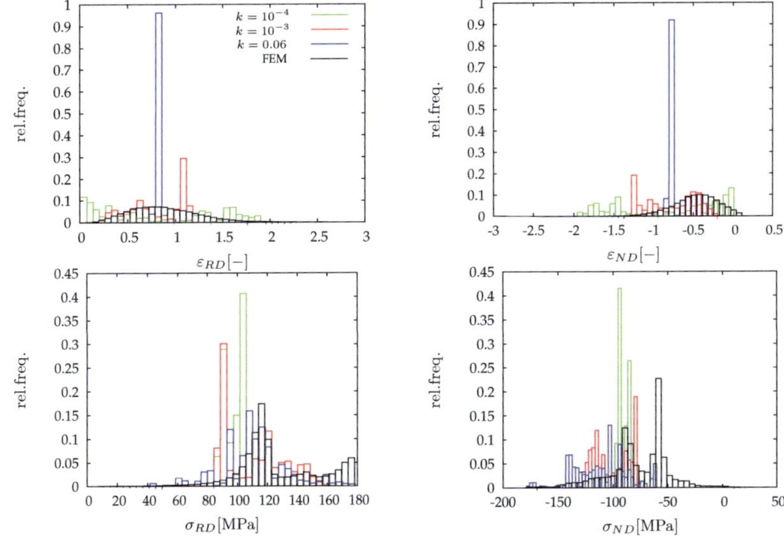

Figure 7.11: Histograms of the (pseudo-)strain and stress components in RD and ND obtained by FEM compared to predictions by the non-linear HS scheme for the AMS method and different values of k at 55% th.red.

Various degrees of thickness reduction

In order to extend the study on the influence of the choice of the comparison medium which is prospectively pretty valuable with regard to the change of the characteristics of rolling textures (e.g., transition from alloy- to pure metal texture, see discussion in (Kocks et al., 1998)), three more degrees of thickness reduction during rolling are considered in the sequel, being 50%, 70% and 90%. Table 7.3 shows the CODFs predicted for an isotropic comparison medium (iso) and two very different factors k. Especially examining the cube component for 50% th.red. and the brass component for 90% th.red., it becomes obvious that the choice of the stiffness of the comparison medium strongly influences the type of the predicted texture. Furthermore, the separate influences of the anisotropy of the comparison medium (A), the incorporation of the ellipsoidal morphology (M) according to the macroscopic deformation ($A \sim \bar{F}$), and the localization of the spin (S; $\omega_\alpha \neq \bar{\omega}$) are investigated for two different factors of k and 70% th.red. in Tab. 7.4. Although, as expected, the effects of A , M and S are not

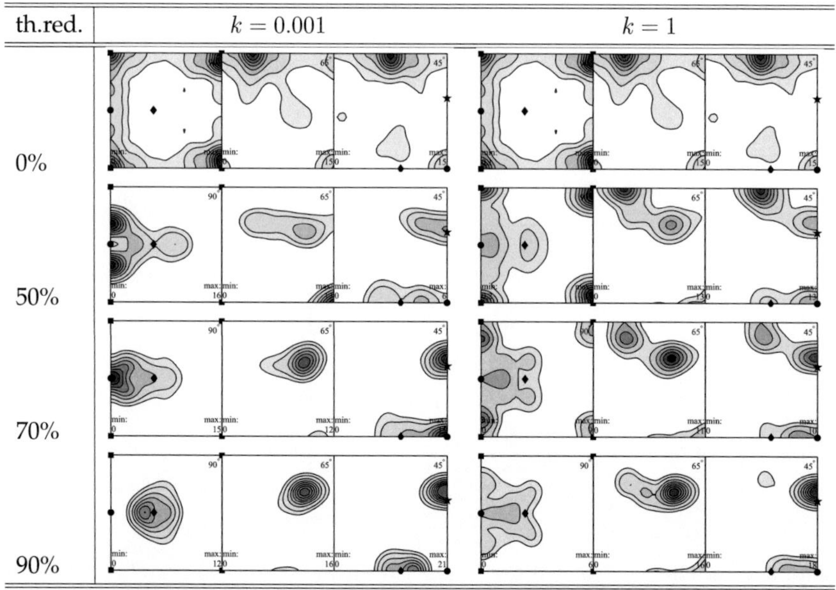

Table 7.3: CODFs for $\mathbb{C}_0 = k\,\mathbb{C}_{\mathrm{Voigt}}^{\mathrm{iso,elast}}$ for different degrees of thickness reduction and two different factors k; characteristic texture components: ● Goss, ★ Copper, ■ Cube, ♦ Brass

as significant as the scaling of the stiffness of the comparison medium by k (cf. Tab. 7.3), they are, however, observable and do result in pronouncing different texture characteristics. For example, changing the comparison stiffness from the isotropic arithmetic mean of the phase stiffnesses to the inherently anisotropic arithmetic mean of the algorithmic tangents of the phases, the Goss component weakens (row 1 to row 2 in Tab. 7.4). If, additionally, the ellipsoidal grain morphology is accounted for (row 3), the brass component weakens, but a peak in between G and B occurs. This component becomes even more pronounced, if heterogeneous crystal spin velocities are permitted showing that in this case, a stronger tendency of rotating towards the B orientation occurs (row 4).

To characterize the anisotropy of rolled sheets, usually the r-value (also Lankford coefficient) is used (see, e.g., Kocks et al. (1998)). In tensile tests with specimen cut from the sheet metal at different angles to RD, the ratio of the strain rates in width and thickness directions defines the r-value. This coefficient

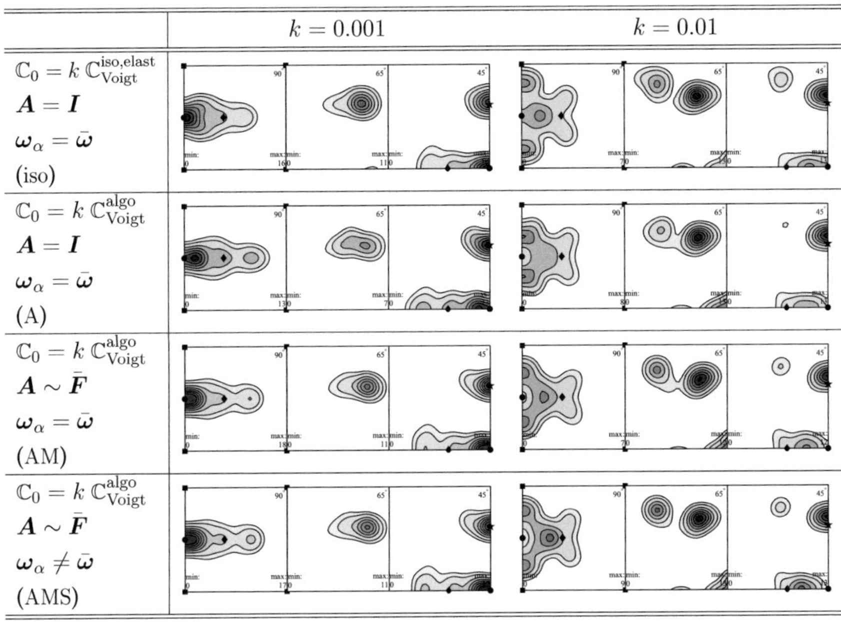

Table 7.4: Influence of $\mathbb{C}_0$, $\boldsymbol{A}$, $\mathbb{P}_0^A$ on the CODFs for 70% th.red. and two different factors of k

can reach values between 0 and ∞, describing no widening and no thinning, respectively. Furthermore, the contraction ratios q, being the inelastic and anisotropic version of Poisson's ratio, are usually evaluated. q_w and q_t are defined as the negative of the width or thickness to length strain rate ratios, respectively. In case of isotropic material behavior, $q_w = q_t = 0.5$ holds. Fig. 7.12 depicts the r-, q_w- and q_t-distribution evaluated with the predicted textures at three different degrees of rolling for three different k in the case of an isotropic comparison material. Obviously, the behavior strongly depends on the chosen comparison stiffness. In all cases, the smallest k predicts the strongest anisotropy of the behavior, which is consistent with, e.g., the findings of M'Guil et al. (2010) also showing that the Taylor-type behavior gives a more isotropic distribution of r-values than the Sachs-like behavior. In thickness direction, the contraction q is mainly larger compared to the one in width direction, whereas there exists also the reverse trend in the case of the softer comparison stiffness

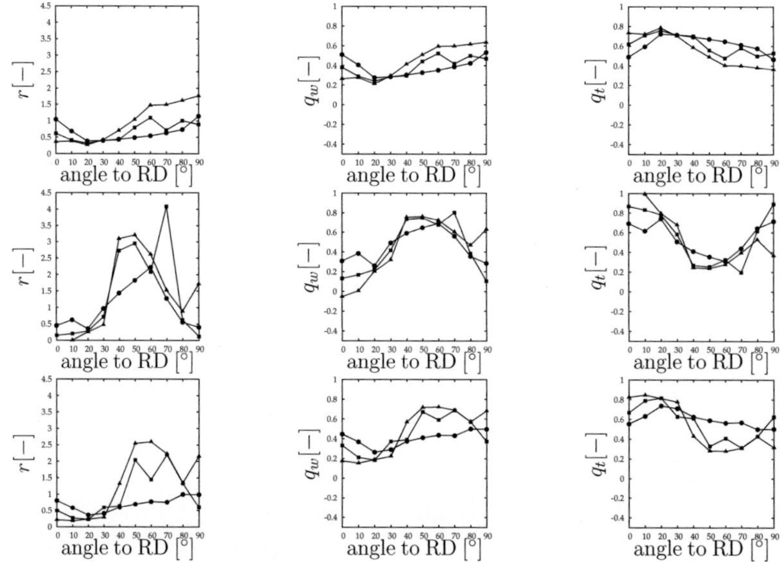

Figure 7.12: r- and q-values ($\blacktriangle$: $k = 0.001$, $\blacksquare$: $k = 0.01$, $\bullet$: $k = 1$) in tensile experiments for different angles to RD and for predicted rolling textures with different thickness reductions (50%, 70%, 90% from top to bottom) in the case of $\mathbb{C}_0 = k\, \mathbb{C}_{\mathrm{Voigt}}^{\mathrm{iso,elast}}$

for angles in between $30°$ and $60°$. Concerning the formability, the non-linear homogenization scheme is able to predict a broad r-range behavior, so that the model offers a potential for obtaining realistic results when simulating complete forming operations with very different types of metals.

7.1.2 Tensile texture

The prescribed deformation in this subsection follows a deviatoric tensile test with the velocity gradient being

$$\boldsymbol{L} = \frac{\dot{\gamma}_0}{\sqrt{6}} \begin{pmatrix} 2 & 0 & 0 \\ 0 & -1 & 0 \\ 0 & 0 & -1 \end{pmatrix}_{\{\boldsymbol{e}_i \otimes \boldsymbol{e}_j\}}. \tag{7.2}$$

Different degrees of elongation are examined extending the RVE in tensile direction by a factor of 2, 3 and 5.

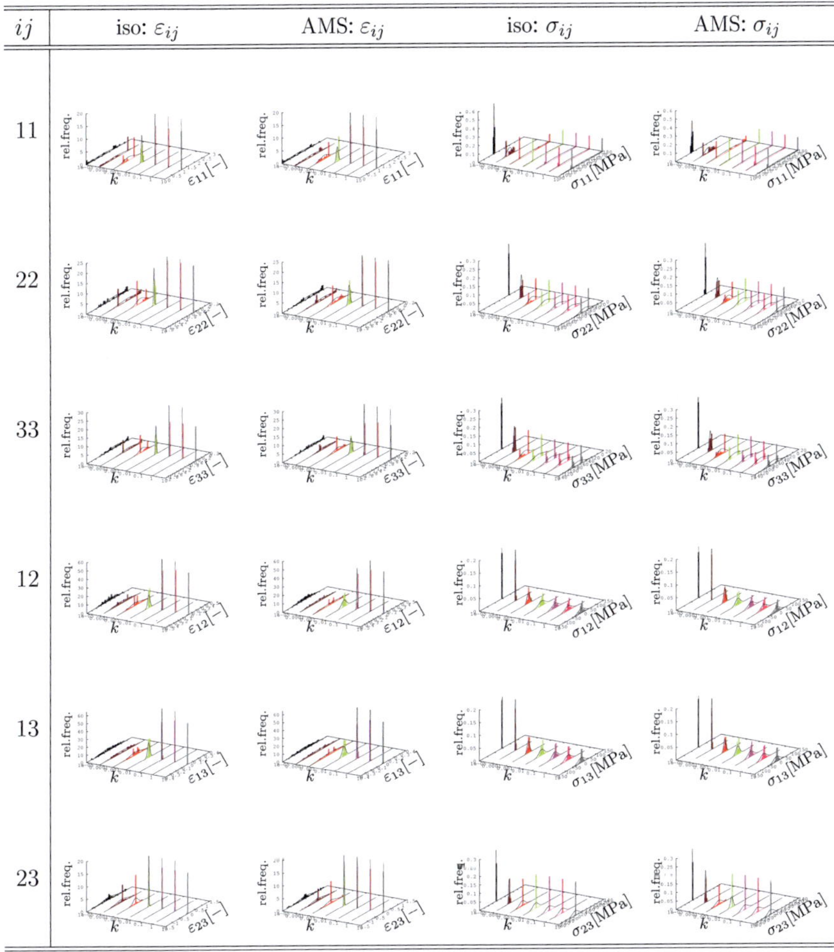

Table 7.5: Pseudo-strain and stress distributions in the volume preserving tensile test in case of 200% elongation computed by the methods iso and AMS

The different degrees of elongation mainly sharpen the texture but do not show a significant difference in the visualization of the CODF, so that all results presented in the sequel for 200% elongation nearly similarly also hold for the other degrees of elongation.

Table 7.5 depicts the stress and pseudo-strain distributions predicted at 200%

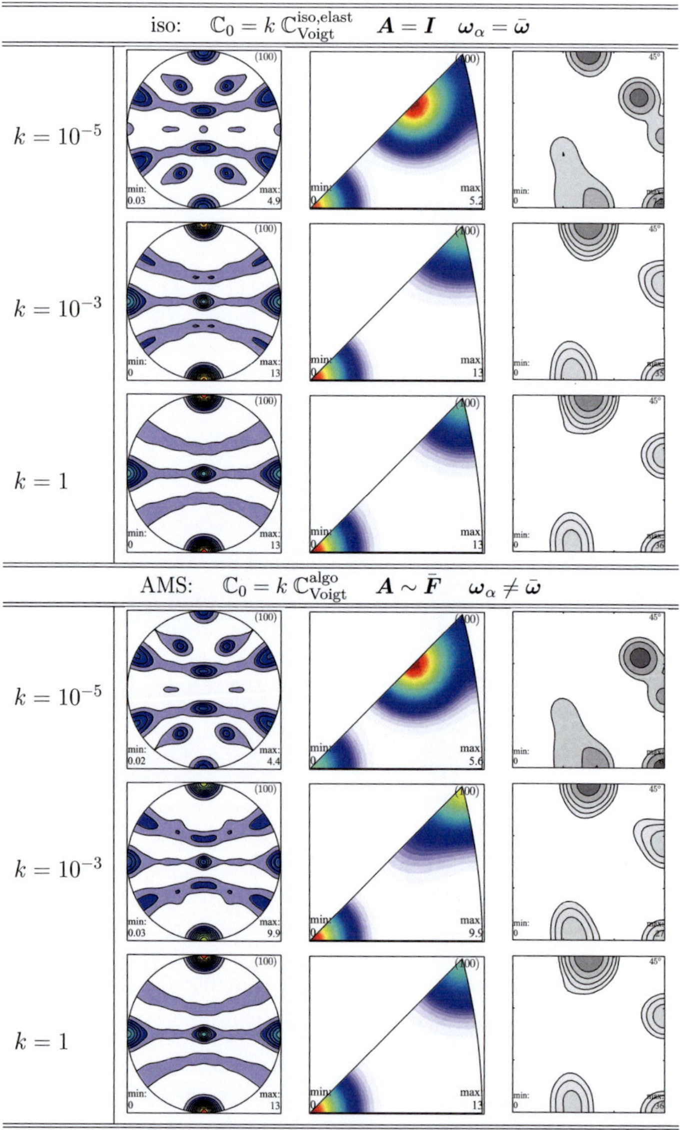

Table 7.6: Tensile textures in terms of (100) pole figures, (100) inverse pole figures and $\varphi_2 = 45°$ slices of the CODFs for the volume preserving tensile test in case of 200% elongation computed by the methods iso and AMS

elongation using 285 texture components and the iso and AMS methods in the NL-HS scheme. Once more, the transition from the homogeneous stress state for small k up to homogeneous strain behavior for large k becomes obvious. While the iso method predicts a homogeneous strain state in all pseudo-strain components for $k \geq 0.1$, the AMS gives a homogeneous strain state for $k \geq 1$. For the smallest of the chosen k being 10^{-5}, the shear stresses show the homogeneous behavior, however, the stress in tensile direction σ_{11} is not perfectly homogeneous. Furthermore, it becomes obvious that in the stress distribution two peaks develop for increasing k. Generally, neither the pseudo-strain nor the stress distribution in the polycrystalline aggregate are predicted to be normally distributed in the tensile experiment.

Table 7.6 shows the corresponding textures accompanying the pseudo-strain and stress states from Table 7.5. One slice of the CODF at $\varphi_2 = 45°$ and the inverse and regular (100) pole figures are depicted for choosing three different comparison stiffnesses in between the Sachs and Taylor behavior. The illustration of inverse pole figures has been chosen for the purpose of comparison with the study of Ahzi and M'Guil (2005). In the inverse pole figure, the G and C components are located at the lower left corner of the triangle, while the Cu component is represented by the upper corner. The brass component is situated on the edge of the triangle approximately at two-thirds on the way from C/G to Cu. The locations of these components in the Euler space have already been shown in Fig. 7.2. Obviously, a significant cube component (as in the initial texture) is present for larger values of k, the intensity of which decreases with decreasing k. For the median among the chosen k, the iso method yields a very strong cube component, while AMS predicts a more balanced ratio between cube and copper. In case of the AMS method, the cube component completely vanishes for the smallest of the chosen k. Both of the methods consistently predict that the copper component rotates somehow towards cube for the smallest k (cf. NDCu in Fig. 7.2) generating a peak in the inverse pole figure near brass and in the surrounding of S, which is also confirmed by the peaks in the CODF.

Qualitatively, the results in Tab. 7.6 agree very well with the results obtained by Ahzi and M'Guil (2005) with their intermediate ϕ-model in between the Sachs and Taylor behavior. However, the method presented here offers a much stronger potential, since the non-linear HS model uses a fourth-order tensor to scale between the extremal behaviors instead of the scalar parameter ϕ in Ahzi's model. It has already been shown for the rolling textures that changing the relation between stress and strain fluctuations from isotropic to anisotropic can result in pronouncing different texture characteristics which, therefore, offers applicability in a much broader domain.

7.1.3 Shear texture

Although textures due to simple shear are not treated very frequently in literature, they are examined in this work to complete the study on deformation textures since they represent a fundamental deformation mode involving rigid body rotation and elongation. Canova et al. (1984) characterized the main texture components of fcc shear textures (resulting from torsion experiments) to be the partial A- and B-fibers as well as the so-called C-component depicted in terms of pole figures in Fig. 7.13. Kocks et al. (1998) state that, additionally, a D-fiber exists being a rotated A-fiber about the shear plane normal. Shear textures are usually not sharp, since each grain is permanently rotating not reaching stable orientations. The visualized textures, therefore, represent only quasi-stationary locations (see, e.g., Kocks et al. (1998)).

Tab. 7.7 shows the shear textures at a deformation of $\gamma = 2$ predicted by the methods iso and AMS. Although the computed textures are not that well-shaped as measured textures due to the small numbers of included discrete orientations (285 components), it becomes obvious that for larger k, the typical shear textures are obtained with elements of the A- and B-fibers. Decreasing k leads to weakening A and sharpening B. In the present example, a strong B-texture is especially observable for $k = 0.001$ and lower k leading to smearing the B-texture showing even some aspects of the above mentioned D-fiber. The difference in the results of the iso and AMS method is comparatively small. In case of AMS, the transition from A to B occurs already for slightly larger values of k. In general, the obtained shear textures show a strong dependence on the stiffness of the comparison material pronouncing different realistic aspects of shear textures. The used homogenization method, therefore, shows to be applicable in predicting shear (or torsion) textures. However, it is recommendable to use a larger number of texture components in the simulations to receive a more continuous texture representation which is less governed by individual orientations for this comparatively weak texture type.

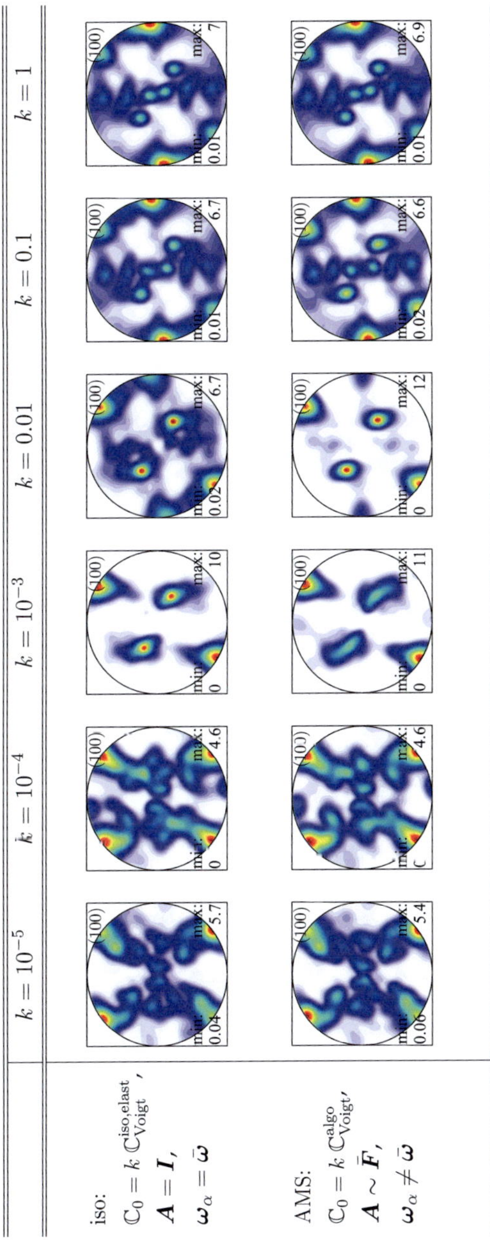

Table 7.7: Shear textures in terms of (100) pole figures for simple shear case of $\gamma = 2$ computed by the methods iso and AMS

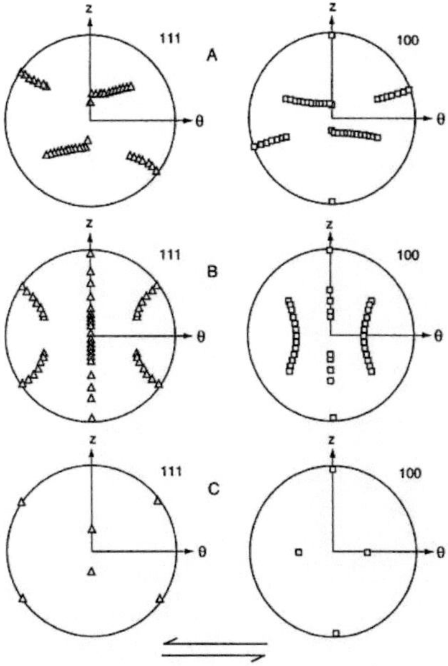

Figure 7.13: Typical shear texture components in the (111) and (100) pole figures (taken from Kocks et al. (1998))

Table 7.8 depicts the corresponding pseudo-strain and stress distributions in the polycrystalline aggregate evoked by the shear deformation. The homogeneous strain state behavior is achieved for $k \geq 0.1$. However, in general, the iso method already yields a more homogeneous state than the AMS method for $k \geq 10^{-2}$, which becomes obvious when examining the most pronounced strain component ε_{12} with non-zero mean value due to the deformation. The homogeneous stress state in σ_{12} is roughly achieved for $k = 10^{-5}$, while the other stress components almost show homogeneity for $k \leq 10^{-4}$. The distributions approach the Gaussian type behavior much more than the ones for the rolling and tensile deformations. Nevertheless, the σ_{12} component still shows one pronounced peak also for small values of k.

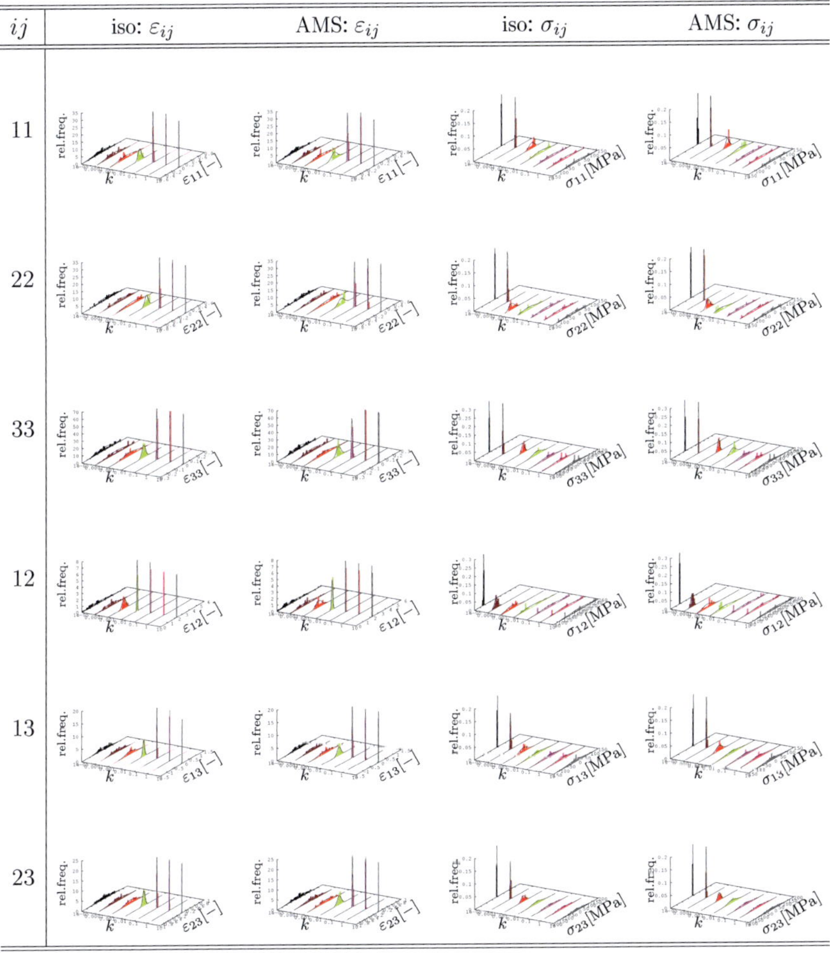

Table 7.8: Pseudo-strain and stress distributions in the shear test in case of $\gamma = 2$ computed by the methods iso and AMS

7.2 Two-scale simulations

This section aims to present prospective applications of the non-linear homogenization scheme in terms of simulations on the macroscale. In order to account for microstructural aspects to study the behavior of whole components, a homogenization method is applied at the integration points of finite element models. The most accurate but also most computationally costly ansatz is to apply a spatially resolved finite element model of the microstructural RVE at the integration point so that the FEM is applied on both scales. Since the computational costs in this case are wholly unacceptable especially in industrial applications, homogenization methods offer a more effective, however, just as well less accurate possibility of accounting for the microstructure of the structural component. Since the non-linear Hashin-Shtrikman-type scheme is pretty flexible and showed to yield very good results, e.g., in texture prediction, it is applied in the sequel in two-scale simulations. For fast computations, the comparison material in the homogenization scheme is chosen to be isotropic (method iso as above), so that $\mathbb{P}_0$ is given analytically when assuming (probably unrealistic) spherical grain morphology.

7.2.1 Tensile tests

As a first example, a tensile experiment is simulated. Exemplarily, the material considered is once more the aluminum alloy AA3104, where the orientation data set of the rolled material, which is usually applied in beverage can production, has been provided by the company Constellium within the scope of project T-A2 of the Research Training Group 1483. The orientation data has been reduced by the method introduced in section 4.1 to a small number of approximately 100 texture components. Due to the geometrical symmetry of the specimen and the orthotropic sample symmetry of the rolled material being aligned with the specimen symmetry, only one eighth of the specimen needs to be modeled and discretized with finite elements applying symmetry boundary conditions. Nevertheless, in the sequel, figures of the complete specimen are shown being generated by mirroring within post processing.

Fig. 7.14 shows the model of the tensile specimen highlighting the elements at which the crystallographic texture will be evaluated. These are two elements being connected to the symmetry plane with the tensile direction as normal (central element, edge element), and one element at the transition zone, where the radius to the larger cross section begins. During deformation, different stress states will be present at these locations resulting in different texture types. Furthermore, Fig. 7.14 exemplarily shows the von Mises stress distribution in the specimen being elongated by 10% and 23%, whereas the latter state already

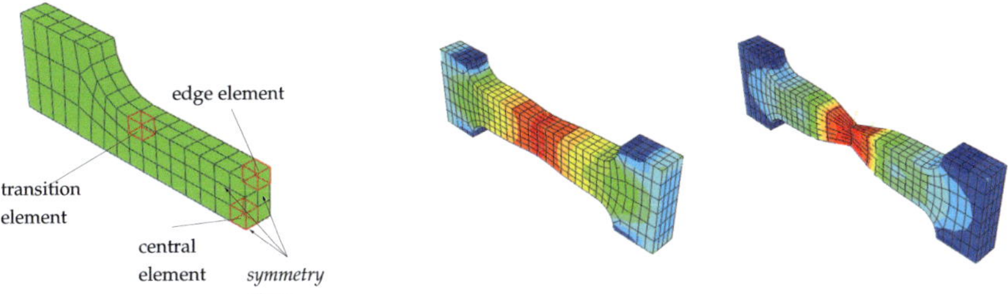

Figure 7.14: Tensile specimen in two-scale computation: positions for texture evaluation (left) and von Mises stress distribution for specimens elongated by 10% and 23% (middle and right)

shows severe necking. The coarse mesh certainly has an impact on this effect, however, since the objective of this study is dedicated to evaluating the influence of the chosen comparison material in the non-linear Hashin-Shtrikman scheme on the results of two-scale simulations rather than doing a mesh convergence study, the mesh is chosen as coarse as possible and has not been varied or adapted.

The tensile test has been simulated using four different factors k governing the stiffness of the isotropic comparison medium ($\mathbb{C}_0 = k\mathbb{C}_{\text{Voigt}}^{\text{iso,elast}}$). Table 7.9 shows the local pseudo-strain and stress distributions in tensile direction evaluated at the three aforementioned locations in the tensile specimen for two different elongations. A homogeneous strain state is approximately predicted in case of $k = 0.5$ for both examined deformation states. Contrarily, the stress state is inhomogeneous for the whole range of considered values of k. However, the increasing stress homogeneity with decreasing k becomes obvious, especially in case of 23% elongation at the transition element. At this large deformation, a behavior that is almost comparison stiffness independent is predicted at the central element showing that probably all grains have already been reoriented to a stable position. Generally, the behavior of the central element and edge element nearly correspond to each other, especially for small and moderate deformations. For the sake of completeness, the distributions of the other pseudo-strain and stress components at these three locations are given in appendix A.3.

The (100) pole figures showing the textures at the three different locations in the specimen are depicted in Table 7.10 for the two different elongations. For

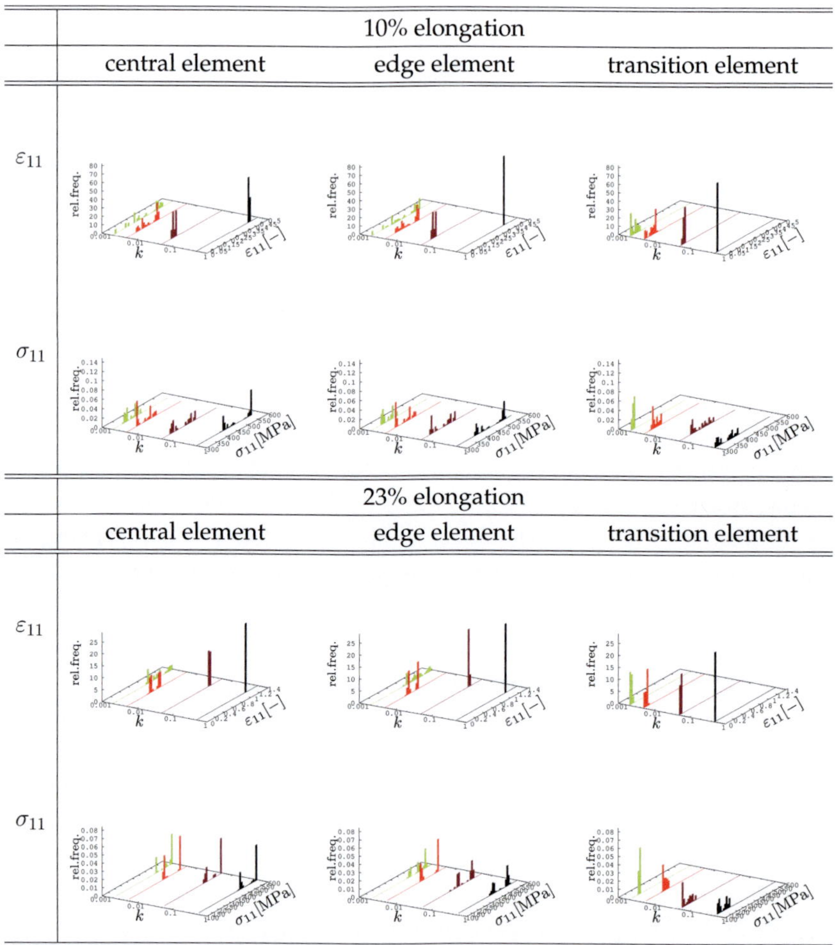

Table 7.9: Pseudo-strain and stress distributions in tensile direction at three different locations of the tensile specimen for 10% and 23% elongation predicted by the two-scale model in case of different factors k

10% elongation, the pole figures do not differ significantly at the three locations, especially the textures at the central and edge elements are approximately equal. However, a variation in the textures with varying the stiffness of the comparison medium is observable. The stronger deformation of 23% shows to result in

144

	10% elongation			23% elongation		
	central element	edge element	transition element	central element	edge element	transition element
$k = 0.5$						
$k = 0.05$						
$k = 0.005$						
$k = 0.002$						

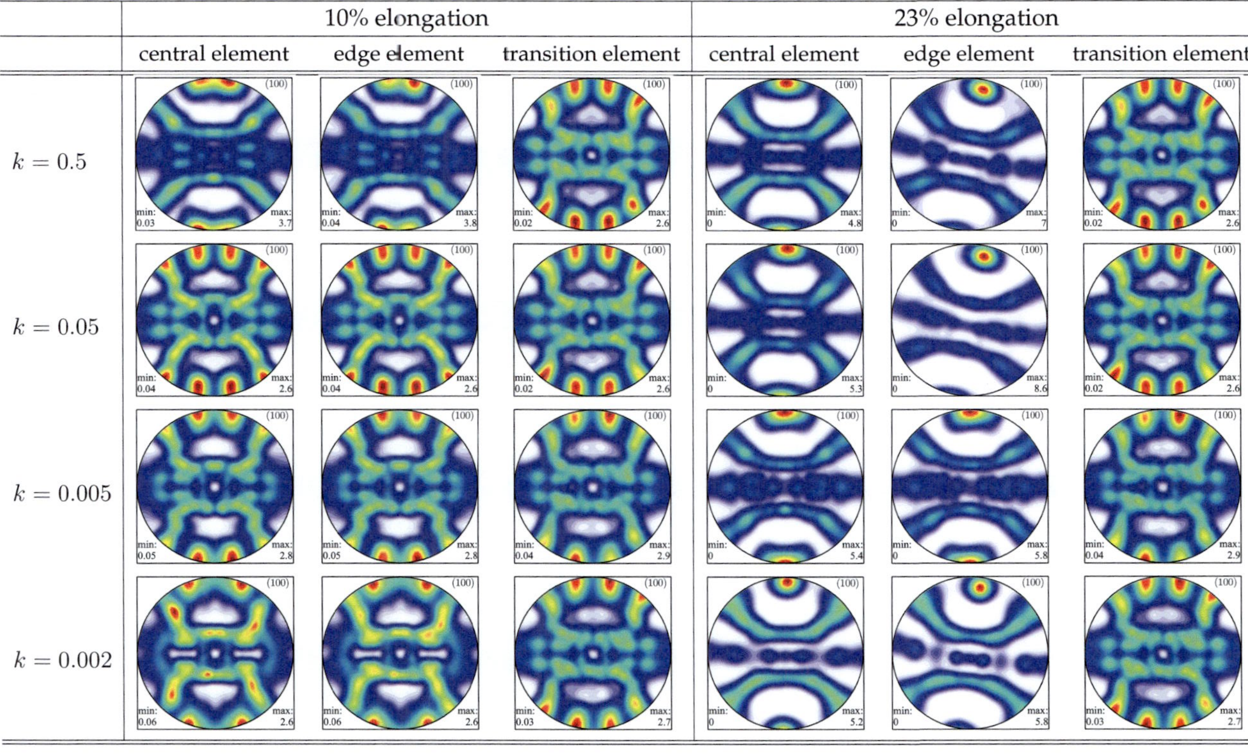

Table 7.10: Texture prediction at three different locations of the tensile specimen in terms of (100) pole figures using the two-scale model for 10% and 23% elongation

stronger variations in the texture. It becomes obvious, that mainly a rigid-body rotation governs the difference in between the central element and edge element textures being most pronounced for large k. The texture at the transition element barely differs in between the two different degrees of elongation. Once more, the texture is significantly influenced by the chosen comparison medium becoming most apparent for the central element. In general, it can be concluded that the choice of comparison material does not only influence the local texture and stress and strain distributions but also results in differences in the overall deformation of the tensile specimen.

7.2.2 Deep drawing

The second example for two-scale simulations is a deep drawing test, where a sheet of 0.265mm thickness and a blank diameter of 60mm is drawn forming a cup of 34mm diameter. These values are based on a typical BUP (Zwick) test. The material considered is an aluminum alloy with the typical texture components of the 2008-T4, which have been identified by Lege et al. (1989) (see also Böhlke et al. (2006b)). The leading four texture components have been

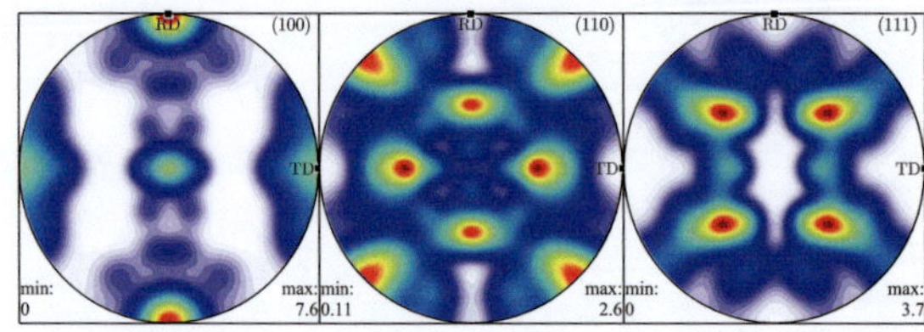

Figure 7.15: Pole figures of the 2008-T4 texture components

orthotropically symmetrized so that 16 texture components are accounted for at each integration point. Fig. 7.15 depicts the pole figures of the 2008-T4 texture generated by the considered texture components. It becomes obvious that the cube component is mainly dominating the texture.

Fig. 7.16 shows the distribution of accumulated plastic slip in the first texture component in the cup for two differently chosen comparison stiffnesses predicted by the homogenization scheme. The accumulated plastic slip shows to be

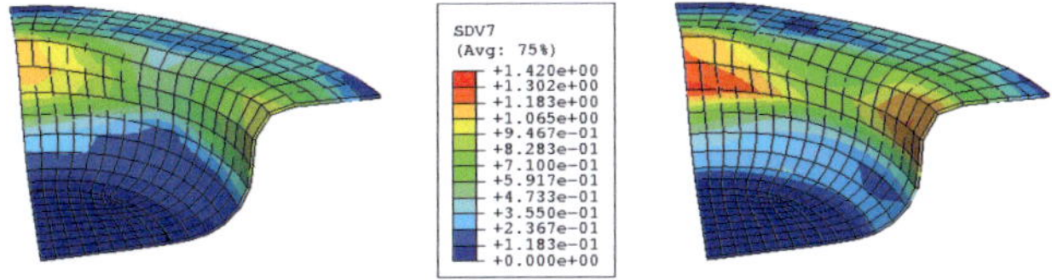

Figure 7.16: Distribution of accumulated plastic slip in the first texture component in a 40% drawn cup for choosing different comparison materials; (left) $k = 0.01$, (right) $k = 1$

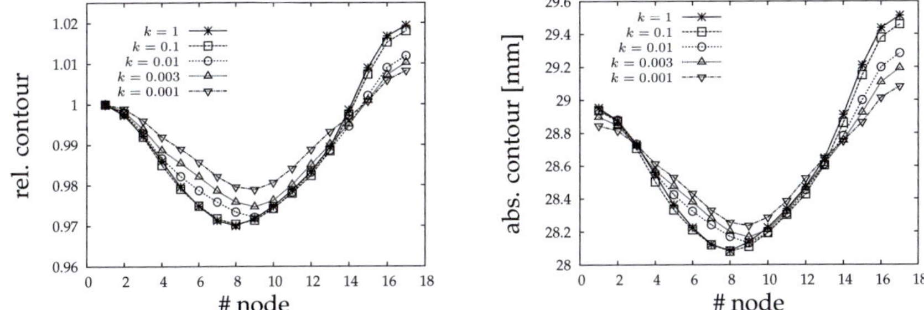

Figure 7.17: Relative and absolute contour of the blank of a 20% drawn cup in case of different comparison stiffnesses

slightly higher for a stiffer comparison medium. Furthermore, the relative and absolute contour of the border of the blank are depicted in Fig. 7.17 for a 20% drawn cup. The contour shows to be strongly influenced by the chosen comparison medium. The shape change of the contour, in the sense of a deviation from a perfect cup with isotropic material behavior, becomes stronger the stiffer the comparison medium is chosen. Decreasing the comparison stiffness leads to slightly shifting the position of the minimum while maintaining the locations of the maxima of the blank shape but the anisotropy of the cup shape (difference between minimum and maximum) decreases. The thickness distribution in one section of the cup, being one of the symmetry sections in the quarter model, is depicted in Fig. 7.18. The thickness of the blank at the bottom of the cup is constant when varying the comparison stiffness (left end of the curves in Fig. 7.18). On the other hand, the thickness of the flange and the undrawn part of the blank show to vary with the comparison stiffness (right end of the curves in

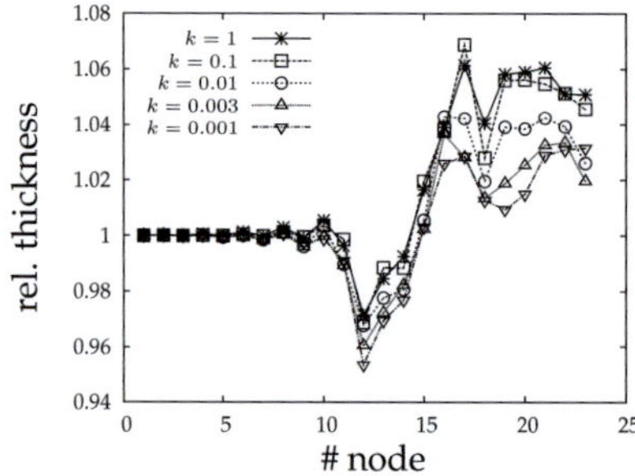

Figure 7.18: Thickness distribution of the blank of a 20% drawn cup in case of different comparison stiffnesses

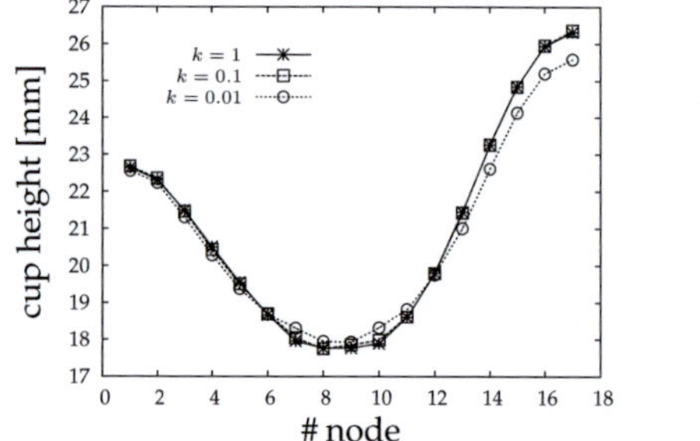

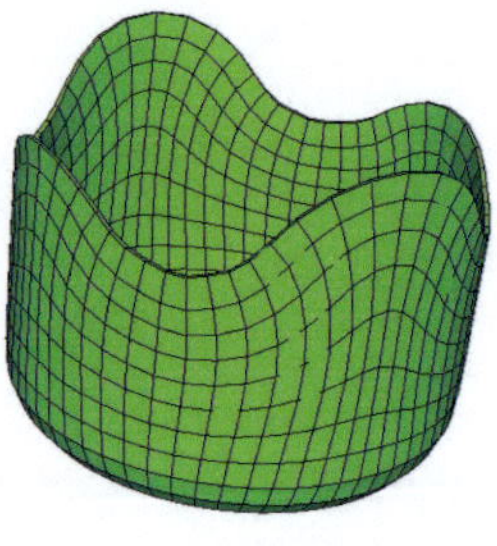

Figure 7.19: Earing profile of the completely drawn cup in case of different comparison stiffnesses (left) and finite element mesh of the drawn cup in case of $k = 0.01$ (right)

Fig. 7.18). The thickening of the undrawn segment is most pronounced in case of the stiff comparison medium. Fig. 7.19 depicts the earing profiles of completely drawn cups in the case of different comparison materials and, exemplarily, the finite element model of the cup in case of $k = 0.01$. Since the computational time increases significantly in case of very compliant comparison media, only three different earing profiles are shown in the diagram. Nevertheless it becomes obvious, that the earing behavior of the completely drawn cup can already be predicted quite well by the contour line of the partly drawn cup (cf. Figs. 7.17 and 7.19).

In summary, strong differences in the behavior of the blank in the deep drawing simulation can be predicted by using the non-linear Hashin-Shtrikman scheme at the integration points to account for the texture. Aiming to compare with real experiments, the proposed two-scale approach offers a tremendous potential for predicting very realistic material response and simultaneously receiving information on the local behavior of the microstructure in complex simulations, e.g., in such forming operations.

Chapter 8

Summary and Conclusions

This thesis examines the prediction of the physically linear and the non-linear behavior of polycrystalline aggregates. Four main topics are highlighted, the first two addressing the representation of microstructures and orientations, and the last two concerning methodical aspects of homogenization.

The first objective is the generation of artificial microstructures. Throughout this work, three-dimensional artificial polycrystalline aggregates are generated for building finite element models. Mainly, the classical Voronoi tessellation is applied using the structured mesh approach due to the requirement of a simple and fast production of hundreds of microstructures for statistical purposes. Additionally, a variant of the classical scheme is proposed. In this case, a classical Poisson-Voronoi mosaic is used as a starting point, the cells of which are expanded or shrunken in order to fulfill a predetermined volume fraction distribution. In an iterative process, finite elements forming the borders of the grains are allowed to cross over to the neighbouring grain. The way of finding the finalized microstructure is not unique since it depends on the sequence of the boundary elements considered. Nevertheless, the obtained microstructures realistically represent real microstructures and are applied within the scope of comparing predictions of homogenization methods with FEM computations.

The second topic addressed in this thesis is the reduction of large experimental orientation data sets generated, e.g., by EBSD or X-ray diffraction measurements. Two methods are proposed, the one being based on the dissection of the orientation space, and the other using the kmeans algorithm in terms of finding orientation clusters. In both cases, an averaging procedure needs to be applied by which one particular orientation is identified being representative for the ones having been averaged. Representing the orientations by quaternions leads to simple rules for finding the mean of several orientations. Two approaches of averaging used in the literature are compared for specific examples. In general, irrespective of the chosen averaging scheme, the dissection method is computationally more efficient compared to the clustering technique. The results of both methods are comparable. While clustering gives better results in case of an a priori knowledge of the numbers of clusters in the texture (pronounced texture components), the dissection method also works well for a gray texture. With the application of both methods, large orientation data sets can be representatively reduced so as to receive a low-dimensional approximation of the orientation distribution function being appropriate for the usage in homogenization schemes. The user needs to decide on the amount of compressing the data so as to receive a sufficiently small data set for subsequent computations, but simultaneously sufficiently large data set to be representative.

The other two main parts of the thesis are devoted to the mechanical linear and non-linear behavior of polycrystals. The objective of the present study is the computationally efficient mechanism-based description of the behavior of microstructured materials. Homogenization schemes are applied on the basis of a homogeneous comparison material and the single inclusion Eshelby solution, which is also reflected in this work. For the physically linear behavior, the self-consistent scheme and the even more efficient singular approximation are applied. While the self-consistent method has an inherently implicit character due to the physically sound choice of the comparison medium being the a priori unknown effective behavior, the singular approximation can be given explicitly. Since the singular approximation uses an isotropic comparison material and assumes negligibility of the nonlocal part of the integral operator, so that, unfortunately, the morphology aspect of the microstructure vanishes, this allows to find an explicit representation of the singular approximation in terms of the fourth-order texture coefficient in case of crystal aggregates with cubic crystal symmetry. Furthermore, the singular approximation includes the Voigt/Reuss bounds and the Hashin-Shtrikman bounds for linear elasticity by appropriately choosing the comparison material. Using the aforementioned approaches, different sources of elastic anisotropy of polycrystals are examined using the examples of oligocrystals and rolled sheets.

The investigations of the physically non-linear behavior of polycrystals in this thesis are based on extensions of the approved homogenization schemes from linear elasticity. On the one hand, the established incremental self-consistent scheme is reflected proposing a modification of using the algorithmic consistent tangents of the grains replacing the continuum moduli, so as to end up with a slighly more compliant solution compared to the original scheme. The incrementally linear solution of non-linear problems is known to yield too stiff results compared to experiments, since the finite discretization of the deformation into steps to be linearized inherently leads to a stiffer response. A new method of non-linearly extending the Hashin-Shtrikman scheme is proposed which is also based on piecewise constant stress polarizations with respect to a (non-linear) homogeneous comparison material. By varying the comparison material, this method shows to offer a great potential in modeling the whole range of admissible behaviors in between the Taylor- and Sachs-type responses of homogeneous strain or stress states. Furthermore, morphological aspects of the microstructure in the ellipsoidal inclusion sense can be added. This method is also applied in the geometrically non-linear regime by updating the lattice spins appropriately. Using the skew-symmetric part of the second derivatives of infinite body's Greens functions, a localization rule for the spin can be derived, which allows to improve the widely-used rule of homogeneous reorientation velocities of the grains. The method is tested extensively for the prediction of texture evolution and is also applied at the integration points of finite elements in order to perform two-scale simulations. The latter framework is - to a great extent - more efficient than FE^2 methods, prospectively also fulfilling the industrial demands for computational efficiency.

In summary, this thesis offers an entire framework for including microstructural information in simulations of structural components. Besides theoretical derivations, the steps of i) preprocessing the experimental data, ii) identification and validation of the model by simulating tensile tests and adopting the appropriate comparison medium by computing texture evolution as well as iii) the integration of the non-linear homogenization scheme at the integration points of a macroscopic finite element model have been presented in detail. Entirely, all these steps in combination represent a new, promising and versatile method to include microstructural data in large-scale simulations involving large deformations and microstructural changes for a detailed computational prediction of real material and structural responses.

Appendix A

A.1 Fibers of AA3104 rolling texture prediction

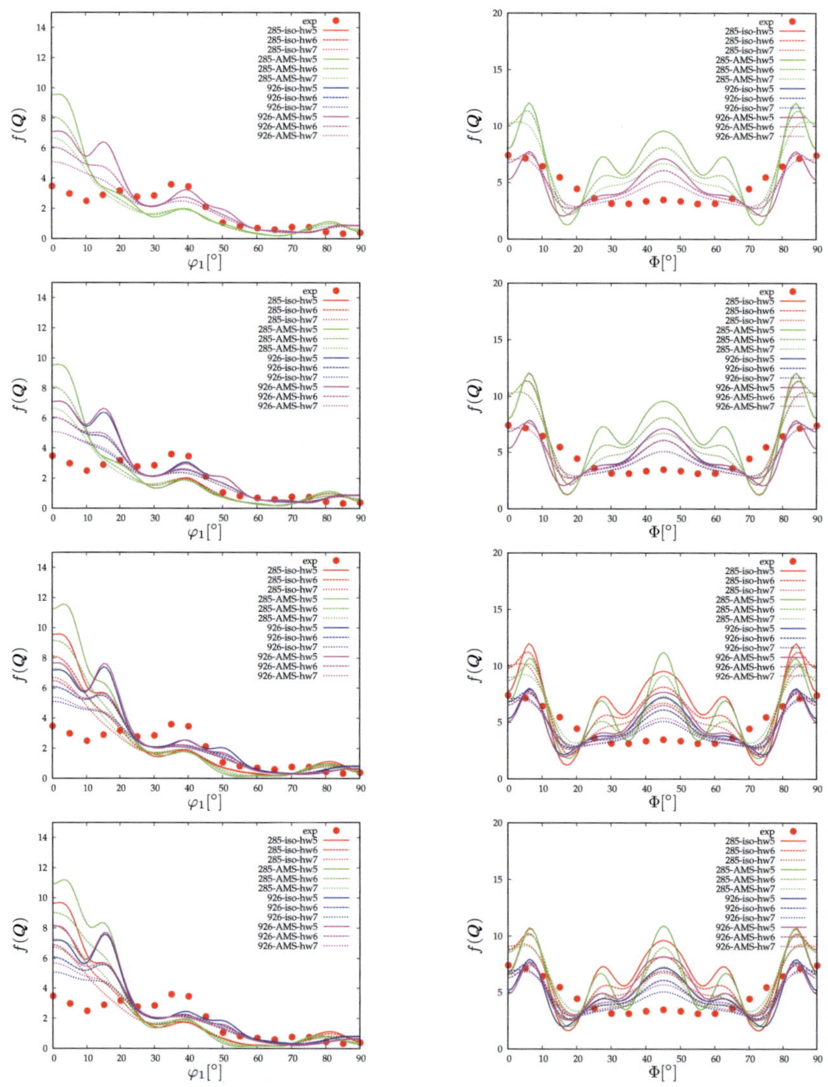

Figure A.1: α- (left) and θ-fiber (right) of approximated CODFs of the texture at 55% th.red.; top to bottom: k=10, 1, 0.1, 0.07

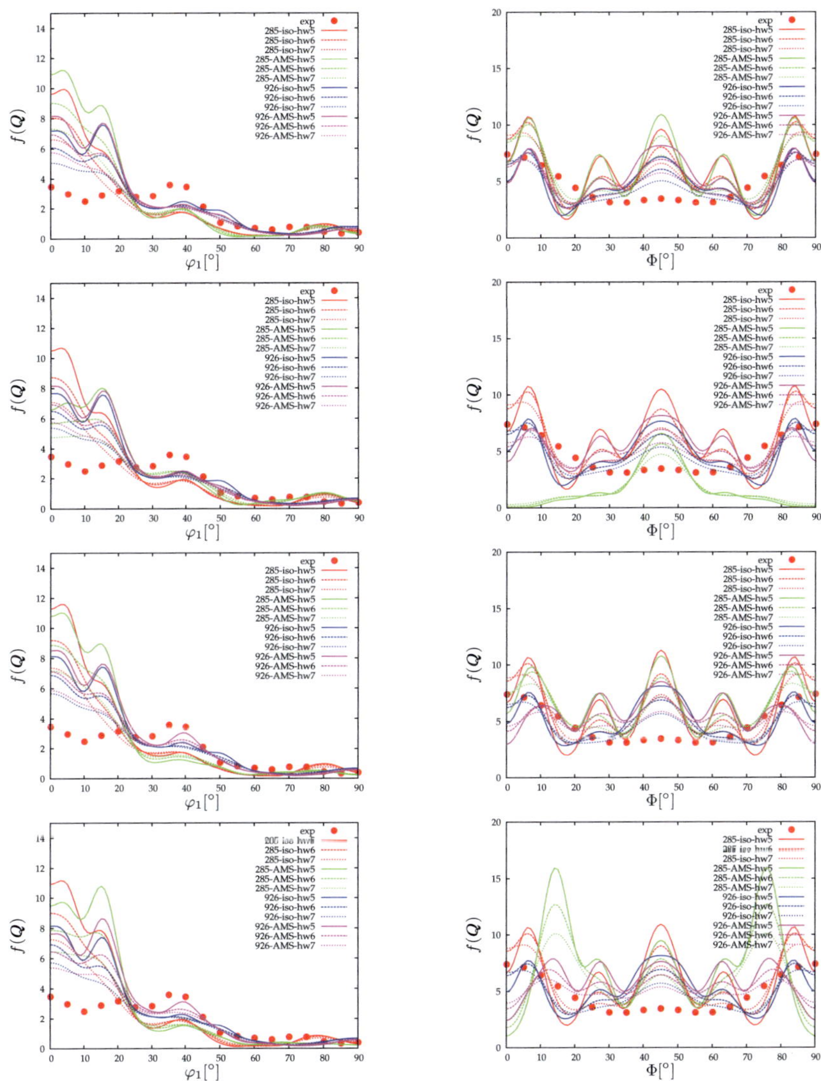

Figure A.2: α- (left) and θ-fiber (right) of approximated CODFs of the texture at 55% th.red.; top to bottom: k=0.06, 0.05, 0.04, 0.03

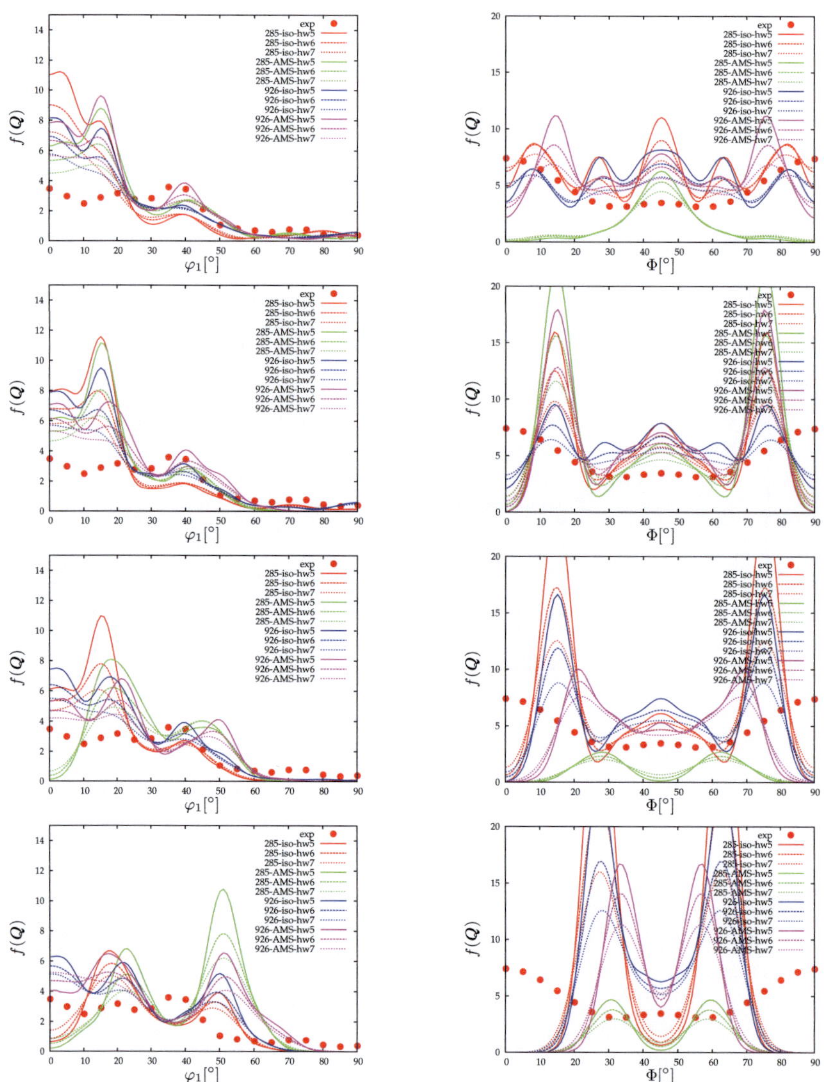

Figure A.3: α- (left) and θ-fiber (right) of approximated CODFs of the texture at 55% th.red.; top to bottom: k=0.02, 0.01, 0.005, 0.002

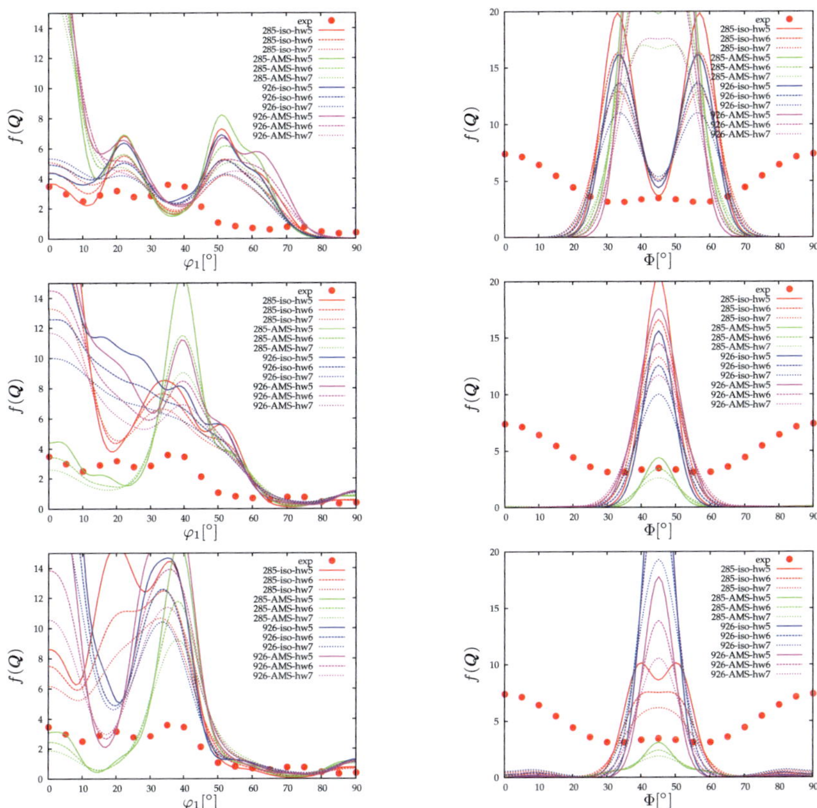

Figure A.4: α- (left) and θ-fiber (right) of approximated CODFs of the texture at 55% th.red.; top to bottom: k=0.001, 0.0001, 0.00001

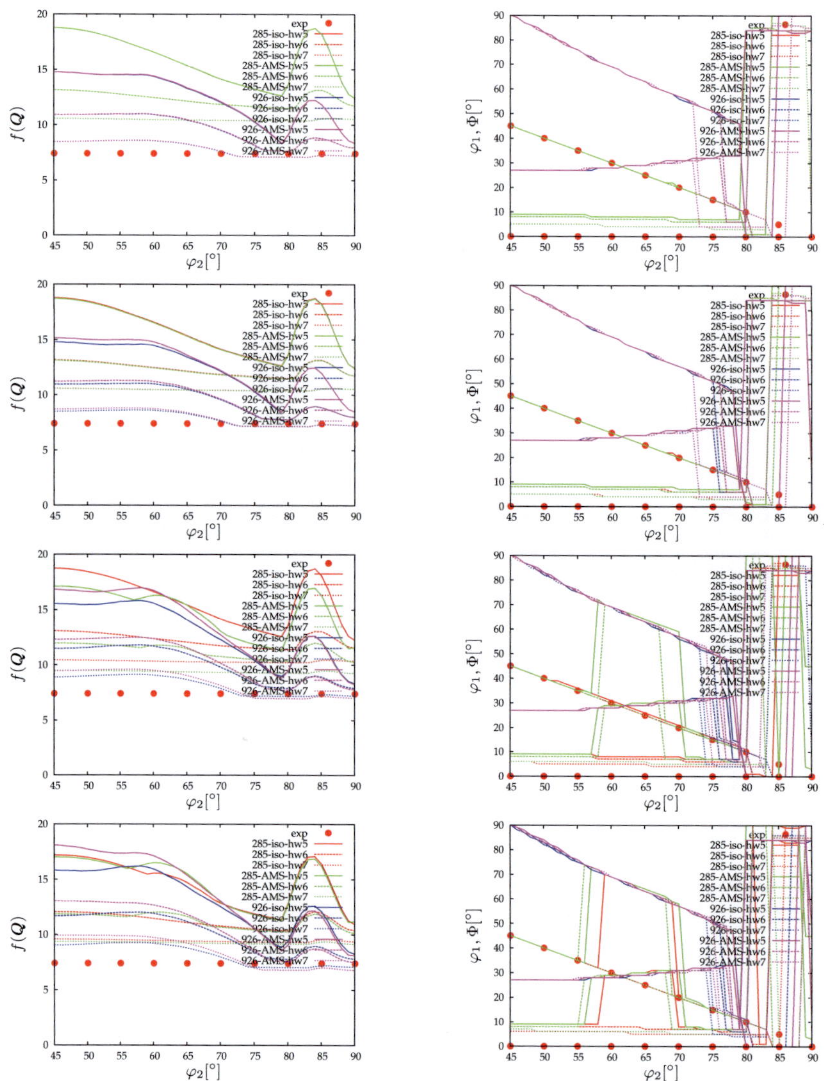

Figure A.5: β-fiber intensity (left) and location (right) of approximated CODFs of the texture at 55% th.red.; top to bottom: k=10, 1, 0.1, 0.07

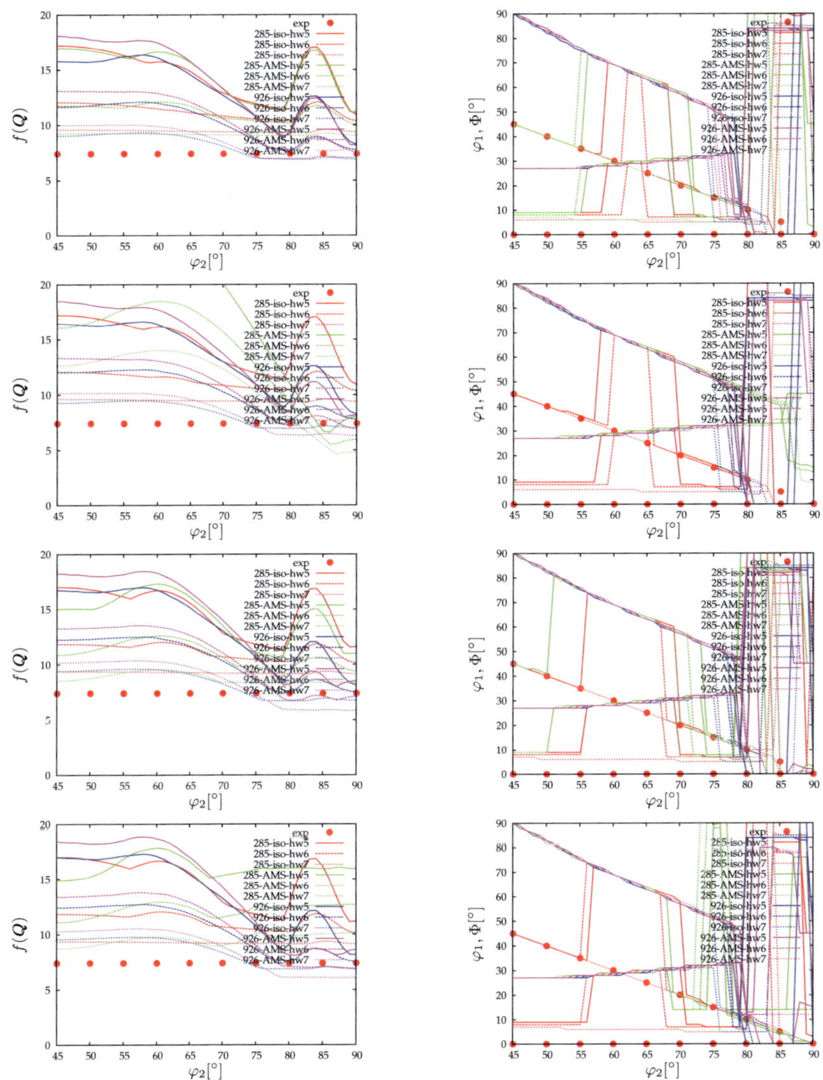

Figure A.6: β-fiber intensity (left) and location (right) of approximated CODFs of the texture at 55% th.red.; top to bottom: k=0.06, 0.05, 0.04, 0.03

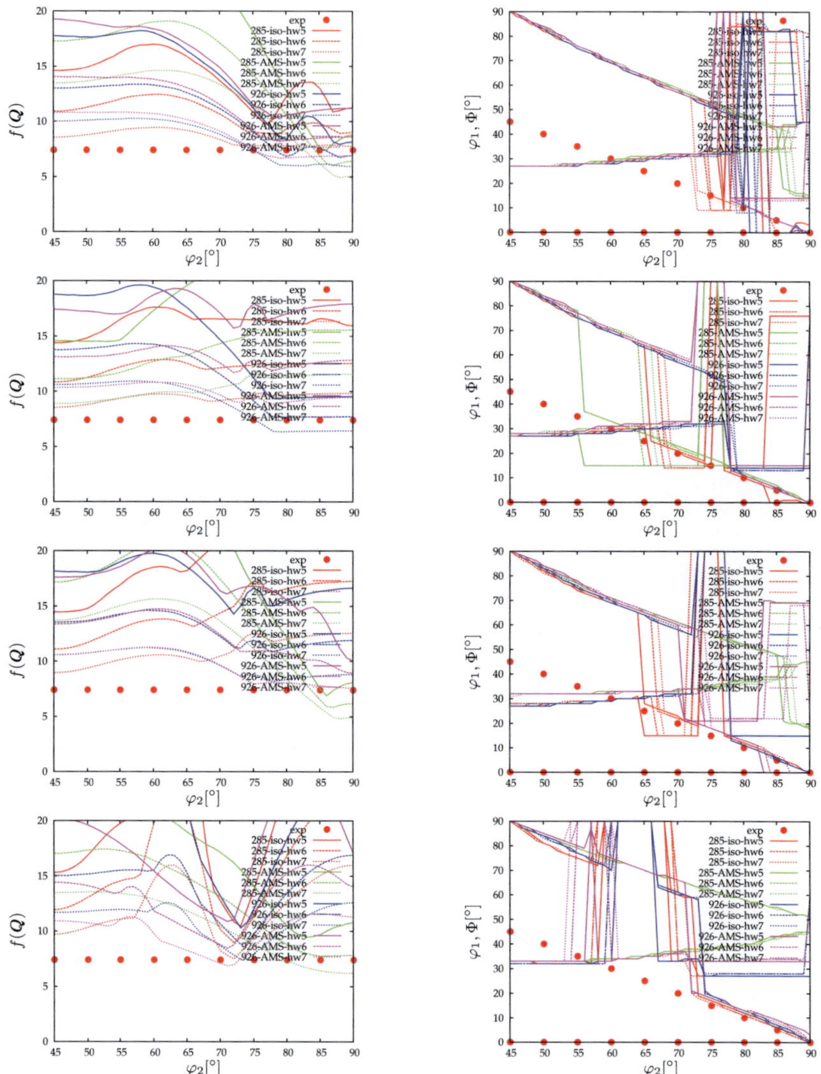

Figure A.7: β-fiber intensity (left) and location (right) of approximated CODFs of the texture at 55% th.red.; top to bottom: k=0.02, 0.01, 0.005, 0.002

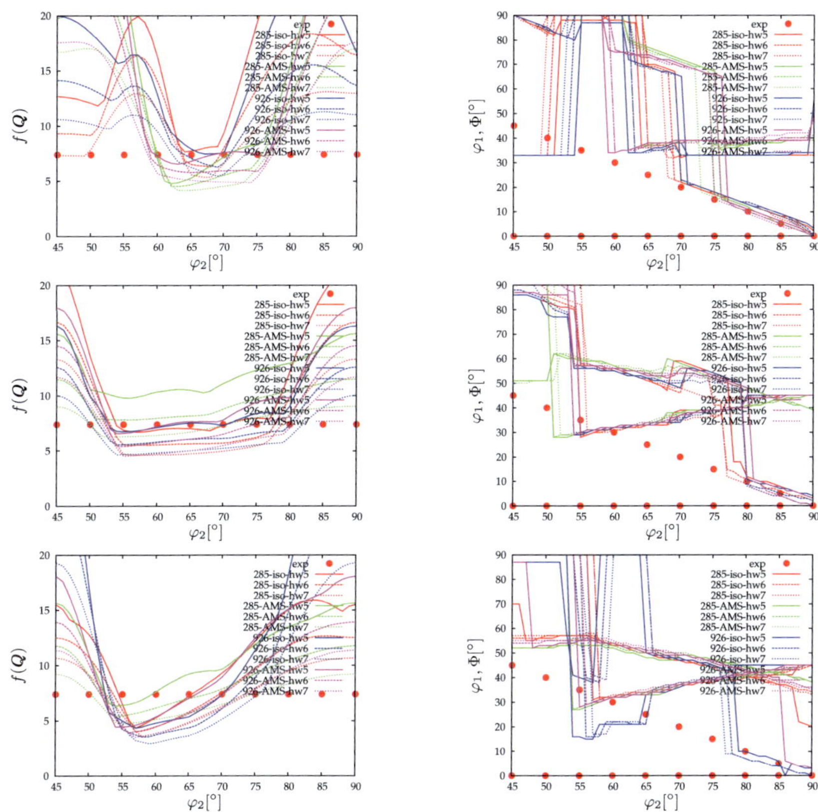

Figure A.8: β-fiber intensity (left) and location (right) of approximated CODFs of the texture at 55% th.red.; top to bottom: k=0.001, 0.0001, 0.00001

A.2 Stress and pseudo-strain histograms of AA3104 rolling texture prediction

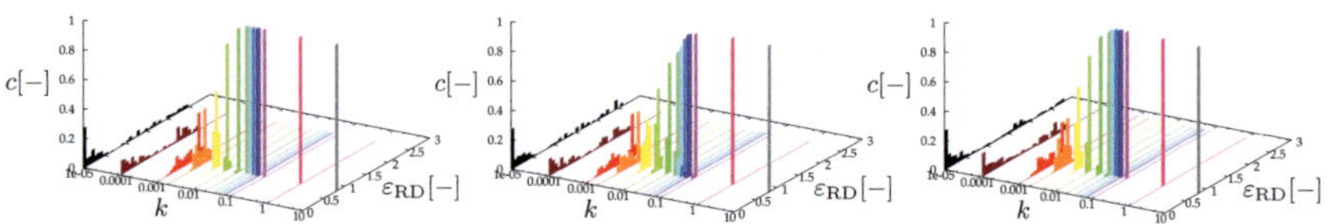

Figure A.9: Pseudo-strain distribution in RD with respect to factor k: 285 iso, 285 AMS, 926 iso (from left to right)

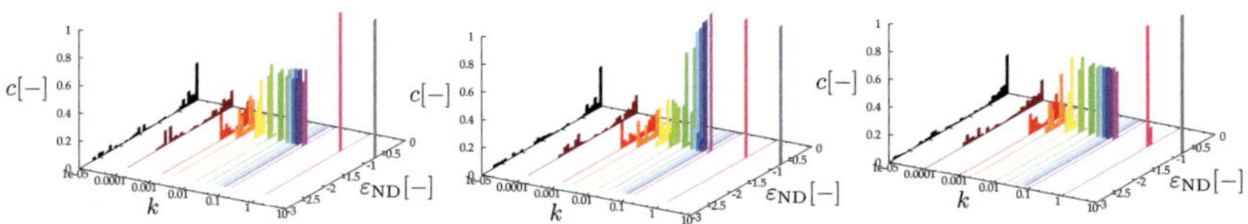

Figure A.10: Pseudo-strain distribution in ND with respect to factor k: 285 iso, 285 AMS, 926 iso (from left to right)

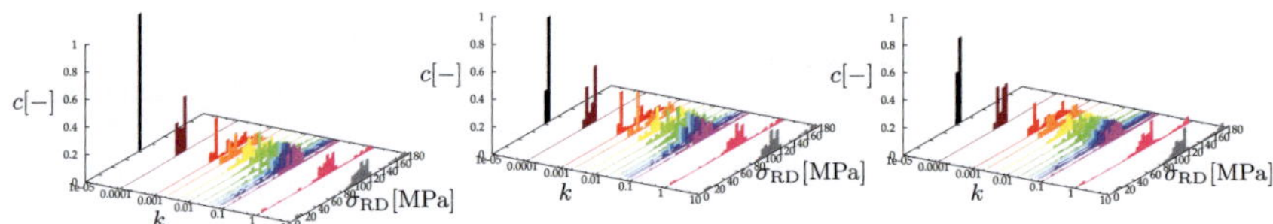

Figure A.11: Stress distribution in RD with respect to factor k: 285 iso, 285 AMS, 926 iso (from left to right)

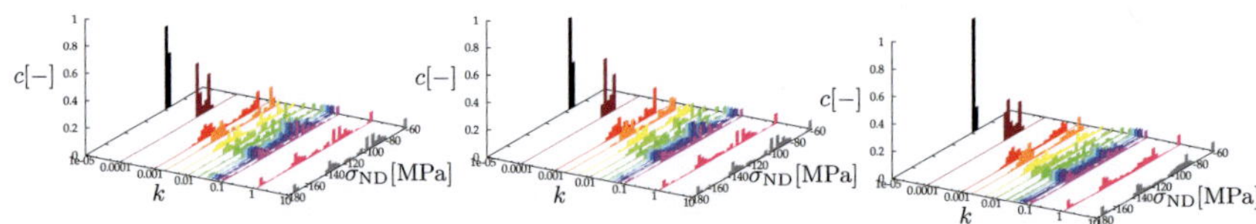

Figure A.12: Stress distribution in ND with respect to factor k: 285 iso, 285 AMS, 926 iso (from left to right)

A.3 Stress and pseudo-strain histograms in two-scale tensile test

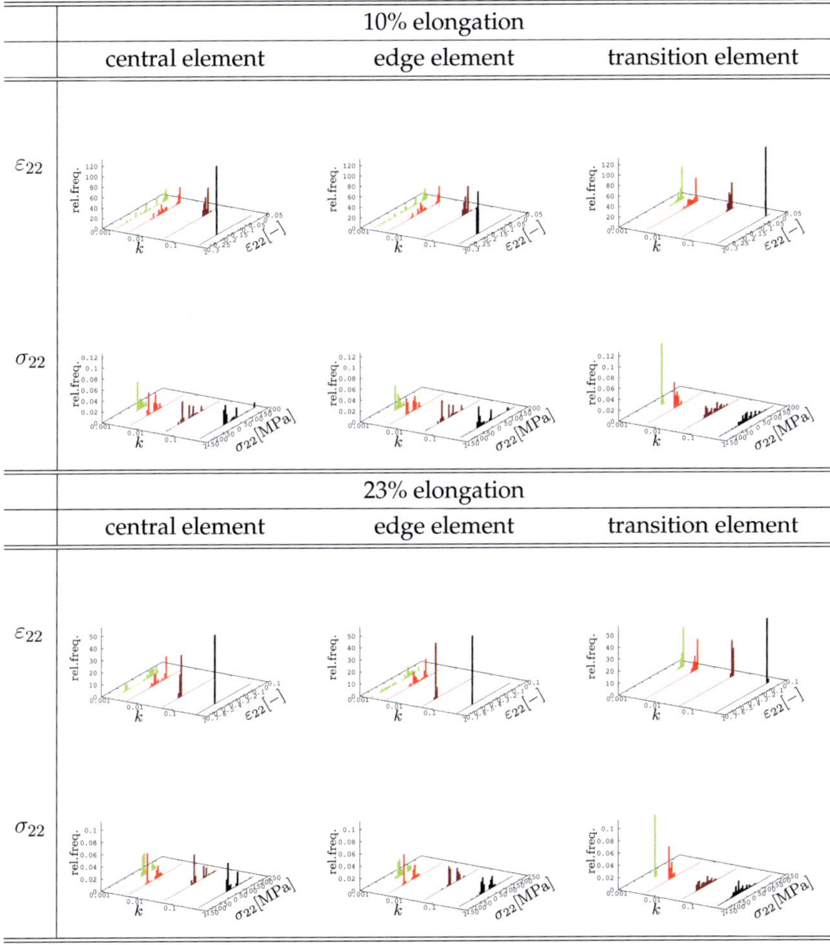

Table A.1: Pseudo-strain and stress distributions in 22-direction at three different locations of the tensile specimen for 10% and 23% elongation predicted by the two-scale model

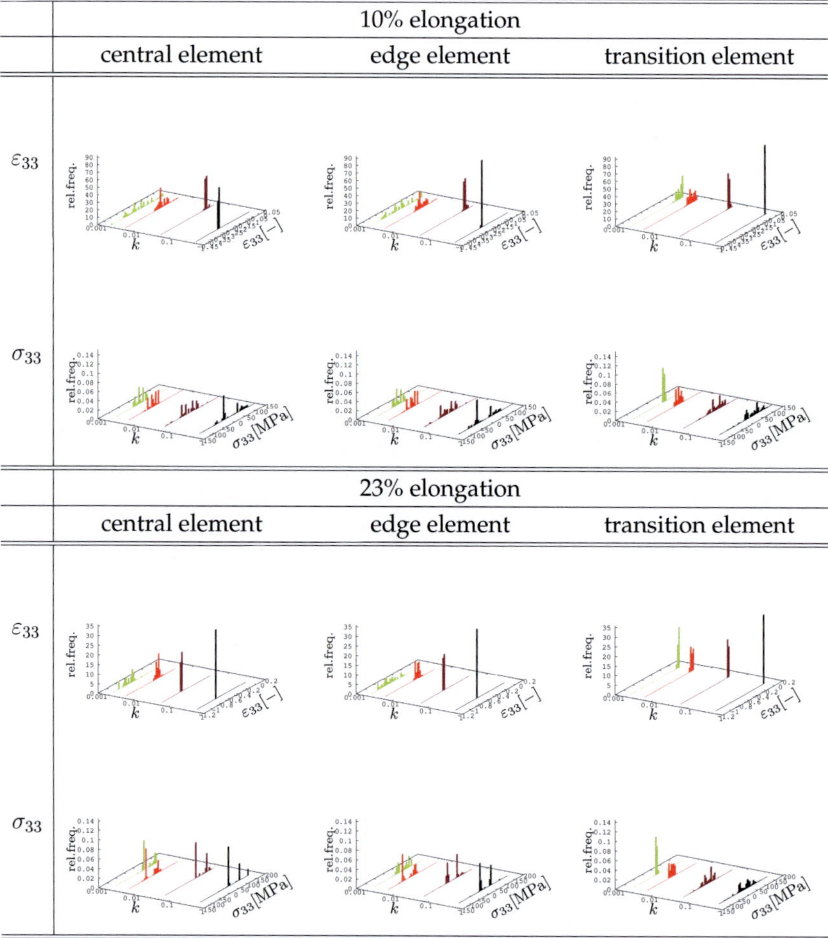

Table A.2: Pseudo-strain and stress distributions in 33-direction at three different locations of the tensile specimen for 10% and 23% elongation predicted by the two-scale model

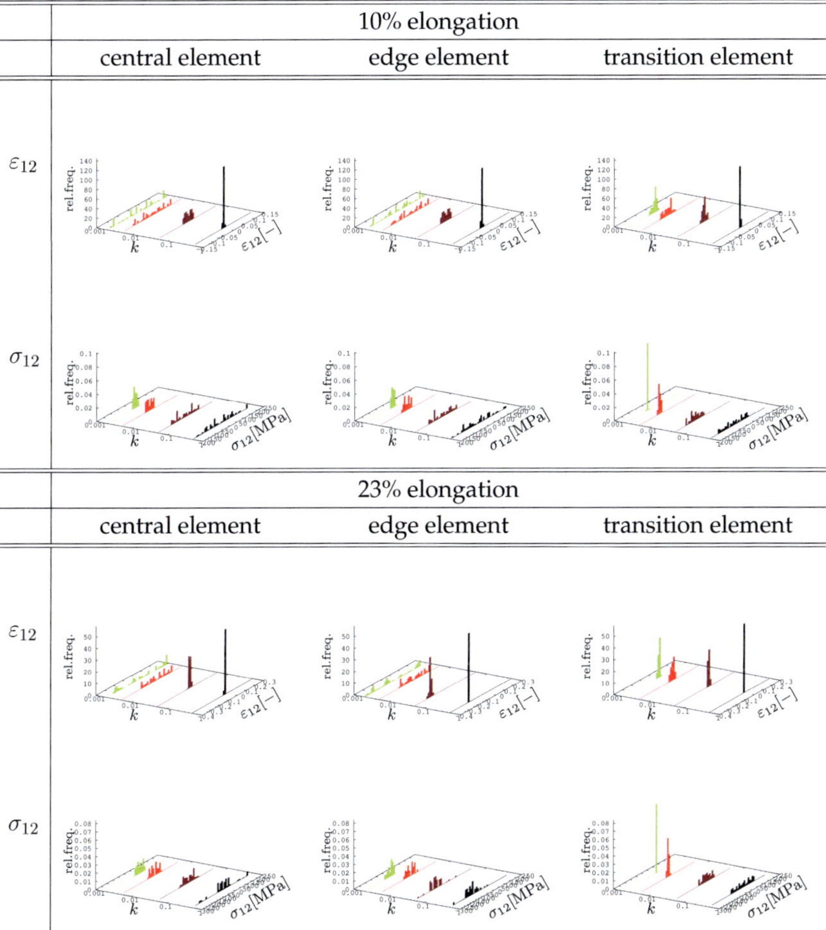

Table A.3: Pseudo-strain and stress distributions in 12-direction at three different locations of the tensile specimen for 10% and 23% elongation predicted by the two-scale model

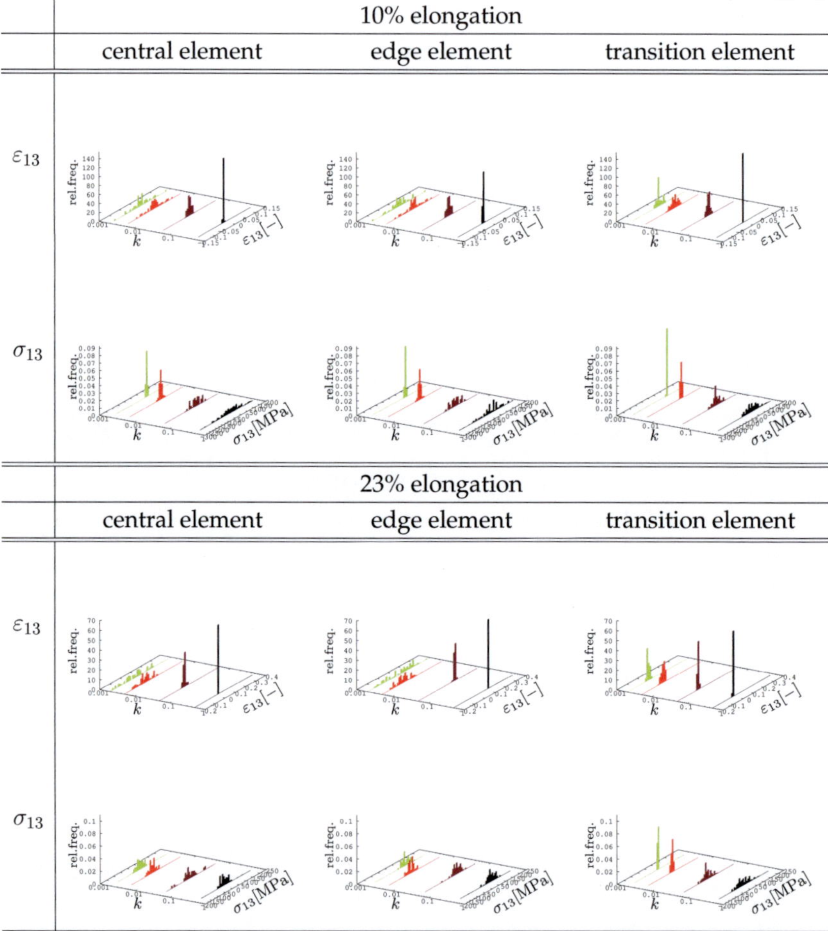

Table A.4: Pseudo-strain and stress distributions in 13-direction at three different locations of the tensile specimen for 10% and 23% elongation predicted by the two-scale model

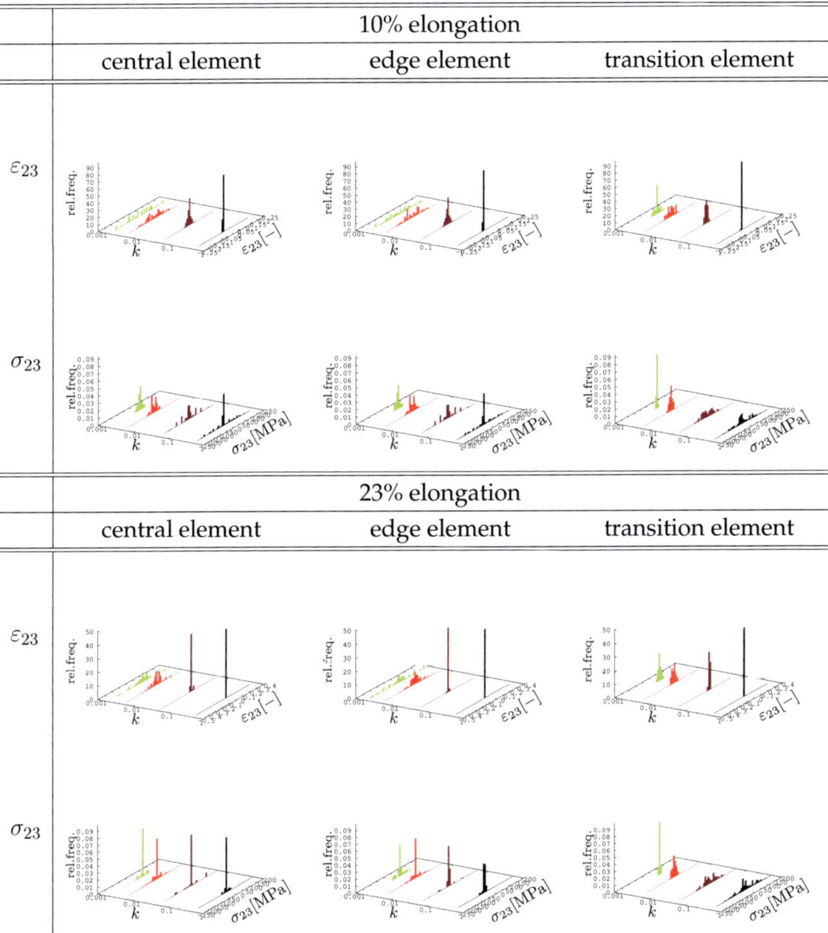

Table A.5: Pseudo-strain and stress distributions in 23-direction at three different locations of the tensile specimen for 10% and 23% elongation predicted by the two-scale model

Bibliography

Adams, B., Boehler, J., Guidi, M., Onat, E., 1992. Group theory and representation of microstructure and mechanical behavior of polycrystals. J. Mech. Phys. Solids 40 (4), 723–737.

Adams, B., Field, D., 1991. A statistical theory of creep in polycrystalline materials. Acta Metall. Mater. 39 (10), 2405 – 2417.

Adams, B., Olson, T., 1998. The mesostructure-properties linkage in polycrystals. Progr. Mat. Sci. 43, 1–88.

Ahzi, S., M'Guil, S., 2005. Simulation of deformation texture evolution using an intermediate model. Solid State Phenomena 105, 251–258.

Ahzi, S., M'Guil, S., 2008. A new intermediate model for polycrystalline viscoplastic deformation and texture evolution. Acta Mater. 56 (19), 5359 – 5369.

Aleksandrov, K., Aisenberg, L., 1966. A method of calculating the physical constants of polycrystalline materials. Soviet Physics - Doklady 11, 323–325.

Altmann, S. L., 1986. Rotations, Quaternions, and Double Groups. Oxford University Press.

Anand, L., 2004. Single-crystal elasto-viscoplasticity: application to texture evolution in polycrystalline metals at large strains. Comp. Meth. Appl. Mech. Eng. 193 (48-51), 5359 – 5383.

Aretz, H., Luce, R., Wolske, M., Kopp, R., Goerdeler, M., Marx, V., Pomana, G., Gottstein, G., 2000. Integration of physically based models into fem and application in simulation of metal forming processes. Modelling and Simulation in Materials Science and Engineering 8 (6), 881.

Arthur, D., Vassilvitskii, S., 2007. k-means++: the advantages of careful seeding. In: Proceedings of the eighteenth annual ACM-SIAM symposium on Discrete algorithms. SODA '07. Society for Industrial and Applied Mathematics, Philadelphia, PA, USA.

Auhorn, M., 2005. Mechanische Eigenschaften urgeformter Mikroproben aus Au58Ag23Cu12Pd5 und ZrO_2. Ph.D. thesis, University of Karlsruhe (TH).

Auhorn, M., Kasanická, B., Beck, T., Schulze, V., Löhe, D., 2006. Mechanical strength and microstructure of Stabilor-G and ZrO_2 microspecimens. Microsyst. Technol. 12, 713–716.

Aurenhammer, F., 1991. Voronoi diagrams – a survey of a fundamental geometric data structure. ACM Computing Surveys 23 (3), 345–405.

Balzani, D., Schröder, J., Brands, D., 2010. FE^2-simulation of micro-heterogeneous steels based on statistically similar RVEs. In: Proceedings of the IUTAM Conference on Variational Concepts with Applications to the Mechanics of Materials. Vol. 21.

Baumeister, G., Ruprecht, R., Hausselt, J., 2004. Microcasting of parts made of metal alloys. Microsyst. Technol. 10 (3), 261–264.

Becker, R., Panchanadeeswaran, S., 1989. Crystal rotations represented as Rodrigues vectors. Texture Microstruct. 10, 167–194.

Benveniste, Y., Dvorak, G. J., Chen, T., 1991. On diagonal and elastic symmetry of the approximate effective stiffness tensor of heterogeneous media. J. Mech. Phys. Solids 39 (7), 927 – 946.

Beran, M., 1965. Use of the vibrational approach to determine bounds for the effective permittivity in random media. Il Nuovo Cimento (1955-1965) 38, 771–782.

Beran, M., McCoy, J., 1970. Mean field variations in a statistical sample of heterogeneous linearly elastic solids. Int. J. Solids Struct. 6 (8), 1035 – 1054.

Berryman, J. G., Apr 2011. Bounds and self-consistent estimates for elastic constants of polycrystals composed of orthorhombics or crystals with higher symmetries. Phys. Rev. E 83, 046130.

Berryman, J. G., 2012. Evaluating bounds and estimators for constants of random polycrystals composed of orthotropic elastic materials. Int. J. Eng. Sci. 58, 11 – 20.

Bertram, A., 2005. Elasticity and Plasticity of Large Deformations. Springer.

Bertram, A., Olschewski, J., 1991. Formulation of anisotropic linear viscoelastic constitutive laws by a projection method. In: Freed, A., Walker, K. (Eds.), High temperature constitutive modelling: Theory and Application. ASME, mD-Vol. 26, AMD-Vol. 121.

Berveiller, M., 2001. Elastoplasticity of metallic polycrystals by the self-consistent model. In: Lemaitre, J. (Ed.), Handbook of Materials Behavior Models. Academic Press, Burlington, pp. 199 – 203.

Berveiller, M., Zaoui, A., 1979. An extension of the self-consistent scheme to plastically flowing polycrystals. J. Mech. Phys. Solids 26, 325–344.

Betten, J., 1993. Kontinuumsmechanik : Elasto-, Plasto- und Kriechmechanik. Springer-Lehrbuch. Springer, Berlin.

Bhandari, Y., Sarkar, S., Groeber, M., Uchic, M., Dimiduk, D., Ghosh, S., 2007. 3d polycrystalline microstructure reconstruction from fib generated serial sections for fe analysis. Comp. Mater. Sci. 41 (2), 222–235.

Bishop, J., Hill, R., 1951. A theory of the plastic distortion of a polycrystalline aggregate under combined stresses. Phil. Mag. 42, 414–427.

Böhlke, T., 2005. Application of the maximum entropy method in texture analysis. Comp. Mater. Sci. 32, 276–283.

Böhlke, T., 2006. Texture simulation based on tensorial fourier coefficients. Comp. Struct. 84, 1086–1094.

Böhlke, T., Bertram, A., 2001a. Bounds for the geometric mean of 4th-order elasticity tensors with cubic symmetry. Z. angew. Math. Mech. 81 (S2), S333–S334.

Böhlke, T., Bertram, A., 2001b. The evolution of Hooke's law due to texture development in polycrystals. Int. J. Solids Struct. 38 (52), 9437–9459.

Böhlke, T., Fritzen, F., Jöchen, K., Tsotsova, R., 2009. Numerical methods for the quantification of the mechanical properties of crystal aggregateswith morphologic and crystallographic texture. Int. J. Mater. Form. 2, 915–917.

Böhlke, T., Haus, U.-U., Schulze, V., 2006a. Crystallographic texture approximation by quadratic programming. Acta Mater. 54, 1359–1368.

Böhlke, T., Jöchen, K., Kraft, O., Löhe, D., Schulze, V., 2010. Elastic properties of polycrystalline microcomponents. Mechanics of Materials 42 (1), 11 – 23.

Böhlke, T., Jöchen, K., Löhe, D., Schulze, V., 2008. Estimation of mechanical properties of polycrystalline microcomponents. Int. J. Mater. Form. 1, 447–450.

Böhlke, T., Risy, G., Bertram, A., 2006b. Finite element simulation of metal forming operations with texture based material models. Modelling Simul. Mater. Sci. Eng. 14, 365–387.

Brenner, R., Castelnau, O., Gilormini, P., 2001. A modified affine theory for the overall properties of nonlinear composites. Comptes Rendus de l'AcadÃ©mie des Sciences - Series IIB - Mechanics 329 (9), 649 – 654.

Budiansky, B., 1965. On the elastic moduli of some heterogeneous materials. J. Mech. Phys. Solids 13 (4), 223 – 227.

Budiansky, B., Wu, T., 1962. Theoretical prediction of plastic strains of polycrystals. In: Proceeding of the 4th US National Congress of Applied Mechanics, Trans. of the ASME.

Bunge, H.-J., 1965. Zur Darstellung allgemeiner Texturen. Z. Metallkde. 56, 872–874.

Bunge, H.-J., 1968a. Die dreidimensionale Orientierungsverteilungsfunktion und Methoden zu ihrer Bestimmung. Kristall und Technik 3, 439–454.

Bunge, H.-J., 1968b. Über die elastischen Konstanten kubischer Materialien mit beliebiger Textur. Kristall und Technik 3, 431–438.

Bunge, H.-J., 1993. Texture Analysis in Material Science. Cuviller Verlag Göttingen.

Bunge, H.-J., Roberts, W., 1969. Orientation distribution, elastic and plastic anisotropy in stabilized steel sheet. J. Appl. Cryst. 2, 116–128.

Bunge, H. J., Tobisch, J., 1972. The texture transition in α-brasses determined by neutron diffraction. J. Appl. Cryst. 5 (1), 27–40.

Canova, G., Kocks, U., Jonas, J., 1984. Theory of torsion texture development. Acta Metall. Mater. 32 (2), 211 – 226.

Carrasco, C., Camurri, C., Garcia, J., Montalba, C., Prat, O., Rojas, D., 2011. Microstructure investigations and mechanical properties of an al-al2o3 mmc produced by semi-solid solidification. Materialwissenschaft und Werkstofftechnik 42 (6), 542–548.

Cho, J.-H., Rollett, A. D., Oh, K. H., 2004. Determination of volume fractions of texture componets with standard distributions in euler space. Metall. and Mater. Trans. A 35 A (1075–1086).

Cowin, S., 1989. Properties of the anisotropic elasticity tensor. Q. J. Mech. appl. Math. 42, 249–266.

Crumbach, M., Pomana, G., Wagner, P., Gottstein, G., 2001. A Taylor-type deformation texture model considering grain interaction and material properties. part i - fundamentals. In: G. Gottstein, D. M. (Ed.), Recrystallisation and Grain Growth, Proceedings of the First Joint Conference.

deBotton, G., Ponte Castañeda, P., 1995. Variational estimates for the creep behavior of polycrystals. Proc. R. Soc. London A448, 121–142.

Dederichs, P. H., Zeller, R., 1972. Elastische Konstanten von Vielkristallen. Kernforschungsanlage Jülich. Institut für Festkörperforschung. Bericht. Kernforschungsanlage Jülich.

Dederichs, P. H., Zeller, R., 1973. Variational treatment of the elastic constants of disordered materials. Z. Physik 259, 103–116.

Delannay, L., Jacques, P. J., Kalidindi, S. R., 2006. Finite element modeling of crystal plasticity with grains shaped as truncated octahedrons. Int. J. Plast. 22 (10), 1879 – 1898.

Delannay, L., Kalidindi, S. R., Van Houtte, P., 2002. Quantitative prediction of textures in aluminium cold rolled to moderate strains. Mater. Sci. Eng. A 336 (13), 233–244.

Delannay, L., Kalidindi, S. R., Van Houtte, P., 2003. Erratum to "Quantitative prediction of textures in aluminium cold rolled to moderate strains": [Materials Science and Engineering A 336 (1-2) (2002) 233-244]. Mater. Sci. Eng. A 351 (1-2), 358 – 359.

Delannay, L., Van Houtte, P., Van Bael, A., 1999. New parameter model for texture description in steel sheets. Texture Microstruct. 31 (3), 151–175.

Delincé, M., Bréchet, Y., Embury, J., Geers, M., Jacques, P., Pardoen, T., 2007. Structure-property optimization of ultrafine-grained dual-phase steels using a microstructure-based strain hardening model. Acta Mater. 55 (7), 2337 – 2350.

Döbrich, K., Rau, C., Krill, C., 2004. Quantitative characterization of the three-dimensional microstructure of polycrystalline Al-Sn using X-ray microtomography. Metallurgical and Materials Transactions A 35, 1953–1961.

Doghri, I., Ouaar, A., 2003. Homogenization of two-phase elasto-plastic composite materials and structures: Study of tangent operators, cyclic plasticity and numerical algorithms. Int. J. Solids Struct. 40 (7), 1681 – 1712.

Dvorak, G. J., Benveniste, Y., 1992. On transformation strains and uniform fields in multiphase elastic media. Proc. R. Soc. London A 437 (1900), 291–310.

Eisenlohr, P., Roters, F., 2008. Selecting a set of discrete orientations for accurate texture reconstruction. Comp. Mater. Sci. 42 (4), 670 – 678.

Engler, O., Randle, V., 2009. Introduction to Texture Analysis: Macrotexture, Microtexture, and Orientation Mapping. CRC Press.

Eshelby, J., 1957. The determination of the elastic field of an ellipsoidal inclusion and related problems. Proc. R. Soc. London A241, 376–396.

Feyel, F., 2003. A multilevel finite element method (FE^2) to describe the response of highly non-linear structures using generalized continua. Comp. Meth. Appl. Mech. Eng. 192 (28-30), 3233 – 3244.

Fokin, A. G., 1972. Solution of statistical problems in elasticity theory in the singular approximation. J. Appl. Mech. Tech. Phys. 13, 85–89.

Fokin, A. G., 1973. Singular approximation for the calculation of the elastic properties of reinforced systems. Mech. Compos. Mater. 9 (3), 445–449.

Fritzen, F., Böhlke, T., 2010. Three-dimensional finite element implementation of the nonuniform transformation field analysis. Int. J. Numer. Meth. Eng. 84 (7), 803–829.

Fritzen, F., Böhlke, T., 2011. Nonuniform transformation field analysis of materials with morphological anisotropy. Compos. Sci. Technol. 71 (4), 433 – 442.

Fritzen, F., Böhlke, T., Schnack, E., 2009. Periodic three-dimensional mesh generation for crystalline aggregates based on Voronoi tessellations. Computational Mechanics 43 (5), 701–713.

Gao, X., Przybyla, C., Adams, B., 2006. Methodology for recovering and analyzing two-point pair correlation functions in polycrystalline materials. Metall. Mater. Trans. A 37, 2379–2387.

Gawad, J., Bael, A. V., Eyckens, P., Samaey, G., Van Houtte, P., Roose, D., 2013. Hierarchical multi-scale modeling of texture induced plastic anisotropy in sheet forming. Comp. Mater. Sci. 66, 65 – 83.

Geers, M., Brekelmans, W., Janssen, P., 2006. Size effects in miniaturized polycrystalline fcc samples: Strengthening versus weakening Int. J. Solids Struct. 43, 7304–7321.

Geers, M., Kouznetsova, V., Brekelmans, W., 2010. Multi-scale computational homogenization: Trends and challenges. J. Comput. Appl. Math. 234 (7), 2175 – 2182.

Geiger, M., Kleiner, M., Eckstein, R., Tiesler, N., Engel, U., 2001. Microforming. CIRP Annals - Manufacturing Technology 50, 445–462.

Ghosh, S., Moorthy, S., 1995. Elastic-plastic analysis of arbitrary heterogeneous materials with the voronoi cell finite element method. Comp. Meth. Appl. Mech. Eng. 121 (1-4), 373–409.

Gilormini, P., 1996. A critical evaluation of various nonlinear extensions of the self-consistent model. In: Micromechanics of plasticity and damage of multiphase nmaterials. kluwer Academic Publishers, pp. 67–74.

Gilormini, P., Bréchet, Y., 1999. Syntheses: Mechanical properties of heterogeneous media: Which material for which model? Which model for which material? Modelling Simul. Mater. Sci. Eng. 7, 805–816.

González, C., LLorca, J., 2000. A self-consistent approach to the elasto-plastic behaviour of two-phase materials including damage. J. Mech. Phys. Solids 48 (4), 675 – 692.

Groeber, M., Ghosh, S., Uchic, M. D., Dimiduk, D. M., 2008. A framework for automated analysis and simulation of 3d polycrystalline microstructures. part 2: Synthetic structure generation. Acta Mater. 56 (6), 1274 – 1287.

Gross, D., Seelig, T., 2007. Bruchmechanik: Mit einer Einführung in die Mikromechanik. Springer.

Guidi, M., Adams, B., Onat, E., 1992. Tensorial representation of the orientation distribution function in cubic polycrystals. Texture Microstruct. 19, 147–167.

Hain, M., Wriggers, P., 2008. Computational homogenization of micro-structural damage due to frost in hardened cement paste. Finite Elements in Analysis and Design 44 (5), 233 – 244.

Halmos, P., 1958. Finite-Dimensional Vector Spaces. D. van Nostrand Company, Inc., New York.

Hansen, J., Lücke, K., Pospiech, J., 1978. Tables for Texture Analysis of Cubic Crystals. Springer-Verlag, Berlin.

Hashin, Z., Shtrikman, S., 1962. A variational approach to the theory of the elastic behaviour of polycrystals. J. Mech. Phys. Solids 10, 343–352.

Hashin, Z., Shtrikman, S., 1963. A variational approach to the theory of the elastic behaviour of multiphase materials. J. Mech. Phys. Solids 11 (2), 127 – 140.

Haupt, P., 2000. Continuum Mechanics and Theory of Materials. Springer.

Hazanov, S., Huet, C., 1994. Order relationships for boundary conditions effect in heterogeneous bodies smaller than the representative volume. J. Mech. Phys. Solids 42 (12), 1995–2011.

Hencky, H., 1932. On a simple model explaining the hardening effect in poly-crystalline metals. Journal of Rheology 3 (1), 30–36.

Hershey, A., 1954. The elasticity of an isotropic aggregate of anisotropic crystals. J. Appl. Mech. 21, 236–241.

Hill, R., 1952. The elastic behaviour of a crystalline aggregate. Proc. Phys. Soc. London A 65, 349–354.

Hill, R., 1963. Elastic properties of reinforced solids: Some theoretical principles. J. Mech. Phys. Solids 11 (5), 357 – 372.

Hill, R., 1965a. Continuum micro-mechanics of elastoplastic polycrystals. J. Mech. Phys. Solids 13, 89–101.

Hill, R., 1965b. A self-consistent mechanics of composite materials. J. Mech. Phys. Solids 13 (4), 213 – 222.

Hill, R., 1967. The essential structure of constitutive laws for metal composites and polycrystals. J. Mech. Phys. Solids 15 (2), 79 – 95.

Hirsch, J., Lücke, K., 1988a. Mechanism of deformation and development of rolling textures in polycrystalline f.c.c. metals–I: Description of rolling texture development in homogeneous CuZn alloys. Acta metall. 36 (11), 2863 – 2882.

Hirsch, J., Lücke, K., 1988b. Mechanism of deformation and development of rolling textures in polycrystalline f.c.c. metals-II: Simulation and interpretation of experiments on the basis of Taylor-type theories. Acta metall. 36 (11), 2883–2904.

Hosford, W., Galdos, A., 1990. Lower bound yield locus calculations. Texture Microstruct. 12 (1-3), 89–101.

Hosford, W. F., 1977. On orientation changes accompanying slip and twinning. Texture of Crystalline Solids 2 (3), 175–182.

Huang, M., Man, C.-S., 2008. Explicit bounds of effective stiffness tensors for textured aggregates of cubic crystallites. Math. Mech. Solids 13, 408–430.

Huet, C., 1990. Application of variational concepts to size effects in elastic heterogeneous bodies. J. Mech. Phys. Solids 38(6), 813–841.

Humbert, M., Gey, N., Muller, J., Esling, C., Dec 1996. Determination of a Mean Orientation from a Cloud of Orientations. Application to Electron Back-Scattering Pattern Measurements. J. Appl. Crystallogr. 29 (6), 662–666.

Humbert, M., Gey, N., Muller, J., Esling, C., Jun 1998. Response to Morawiec's (1998) comment on *Determination of a mean orientation from a cloud of orientations. Application to electron back-scattering pattern measurements.* J. Appl. Crystallogr. 31 (3), 485.

Hutchinson, J., 1970. Elastic-plastic behavior of polycrystalline metals and composites. Proc. R. Soc. London, Ser. A 319, 247–272.

Hutchinson, J., 1976. Bounds and self-consistent estimates for creep of polycrystalline materials. Proc. R. Soc. London. A 348, 101–127.

Idiart, M., Ponte Castañeda, P., 2007. Field statistics in nonlinear composites. i. theory. Proc. R. Soc. London A 463 (2077), 183–202.

Jafari, R., Kazeminezhad, M., 2011. Microstructure generation of severely deformed materials using voronoi diagram in laguerre geometry: Full algorithm. Comp. Mater. Sci. 50 (9), 2698 – 2705.

Jayabal, K., Menzel, A., 2012. Voronoi-based three-dimensional polygonal finite elements for electromechanical problems. Computational Materials Science 64, 66 – 70.

Jia, N., Peng, R. L., Wang, Y., Zhao, X., 2011. Self-consistent modeling of rolling textures in an austenitic-ferritic duplex steel. Mater. Sci. Eng. A 528 (10 - 11), 3615 – 3624.

Jiang, T., Shao, J., 2009. On the incremental approach for nonlinear homogenization of composite and influence of isotropization. Comp. Mater. Sci. 46 (2), 447 – 451.

Jöchen, K., Böhlke, T., 2009a. Combination of the incremental self-consistent scheme and the finite element method with application to metal matrix composites. Proceedings of the 6th ICCSM .

Jöchen, K., Böhlke, T., 2009b. Incremental self-consistent approach for the estimation of nonlinear material behavior of metal matrix composites. PAMM 9, 427–428.

Jöchen, K., Böhlke, T., 2009c. Prediction of the elastic properties of polycrystalline microcomponents by numerical homogenization. Advanced Engineering Materials 11, 158–161.

Jöchen, K., Böhlke, T., 2010. Studie zum Einfluss der Morphologie von Partikeln in Zweiphasenmaterialien auf das makroskopische elastische Materialverhalten mittels Homogenisierungsmethoden. Begleitband zur 1. jährlichen Klausurtagung 2010 und CCMSE Begleitband zum 3. jährlichen Symposium 2010 , 85–90.

Jöchen, K., Böhlke, T., 2011. Preprocessing of texture data for an efficient use in homogenization schemes. Proc. 10th Int. Conf. Techn. Plast. (ICTP 2011) , 848–853.

Jöchen, K., Böhlke, T., 2012. Prediction of texture evolution in rolled sheet metals by using homogenization schemes. Key Engineering Materials 504-506, 649–654.

Jöchen, K., Böhlke, T., Fritzen, F., 2010. Influence of the crystallographic and the morphological texture on the elastic properties of fcc crystal aggregates. Solid State Phenomena 160, 83–86.

Kalidindi, S., Bronkhorst, C., Anand, L., 1992. Crystallographic texture evolution in bulk deformation processing of fcc metals. J. Mech. Phys. Solids 40 (3), 537–569.

Kalidindi, S. R., 1998. Incorporation of deformation twinning in crystal plasticity models. J. Mech. Phys. Solids 46 (2), 267 – 290.

Kanit, T., Forest, S., Galliet, I., Mounoury, V., Jeulin, D., 2003. Determination of the size of the representative volume element for random composites: statistical and numerical approach. Int. J. Solids Struct. 40, 3647–3679.

Kanouté, P., Boso, D., Chaboche, J., Schrefler, B., 2009. Multiscale Methods for Composites: A Review. Arch. Comput. Meth. Eng. 16, 31–75.

Kikuchi, S., 1928. Diffraction of cathode rays by mica. Japanese Journal of Physics 5, 83–96.

Kim, J. H., Kim, D., Barlat, F., Lee, M.-G., 2012. Crystal plasticity approach for predicting the Bauschinger effect in dual-phase steels. Mater. Sci. Eng. A 539, 259 – 270.

Ko, W. H., 2007. Trends and frontiers of mems. Sensors and Actuators A: Physical 136, 62–67.

Kocks, U., Chandra, H., 1982. Slip geometry in partially constrained deformation. Acta Metall. Mater. 30 (3), 695 – 709.

Kocks, U., Tome, C., Wenk, H., 1998. Texture and Anisotropy: Preferred Orientations in Polycrystals and Their Effect on Materials Properties. Cambridge Univ. Pr.

Kraska, M., Doig, M., Tikhomirov, D., Raabe, D., Roters, F., 2009. Virtual material testing for stamping simulations based on polycrystal plasticity. Comp. Mater. Sci. 46 (2), 383 – 392.

Kröner, E., 1958. Berechnung der elastischen Konstanten des Vielkristalls aus den Konstanten des Einkristalls. Z. Phys. 151, 504–518.

Kröner, E., 1961. Zur plastischen Verformung des Vielkristalls. Acta. Metall. 9, 155–161.

Kröner, E., 1972. Complete macroscopic description of linear-dielectric heterogeneous materials. physica status solidi (b) 49 (1), K23–K28.

Kröner, E., Apr. 1977. Bounds for effective elastic moduli of disordered materials. J. Mech. Phys. Solids 25, 137–155.

Kumar, S., Kurtz, S., Banavar, J., Sharma, M., 1992. Properties of a three-dimensional Poisson-Voronoi tesselation: A Monte Carlo study. Journal of Statistical Physics 67, 523–551.

Kuramae, H., Ikeya, Y., Sakamoto, H., Morimoto, H., Nakamachi, E., 2010. Multi-scale parallel finite element analyses of ldh sheet formability tests based on crystallographic homogenization method. Int. J. Mech. Sci. 52 (2), 183 – 197.

Lautensack, C., Zuyev, S., 2008. Random Laguerre tessellations. Adv. in Appl. Probab. 40, 630–650.

Lebensohn, R. A., 2001. N-site modeling of a 3D viscoplastic polycrystal using Fast Fourier Transform. Acta Mater. 49 (14), 2723–2737.

Lebensohn, R. A., Leffers, T., 1999. The rules for the lattice rotation accompanying slip as derived from a selfconsistent model. Textures and Microstructures 31, 217–230.

Lebensohn, R. A., Ponte Castañeda, P., Brenner, R., Castelnau, O., 2011. Computational Methods for Microstructure-Property Relationships. Springer (eds. S.Ghosh, D. Dimiduk), Ch. : Full-Field vs Homogenization Methods to Predict Microstructure-Property Relations for Polycrystalline Materials, pp. 393–441.

Lebensohn, R. A., Tomé, C., 1993. A self-consistent anisotropic approach for the simulation of plastic deformation and texture development of polycrystals: Application to zirconium alloys. Acta Metall. Mater. 41 (9), 2611 – 2624.

Lebensohn, R. A., Tomé, C. N., 1994. A self-consistent viscoplastic model: prediction of rolling textures of anisotropic polycrystals. Mater. Sci. Eng. A 175 (1-2), 71 – 82.

Leffers, T., 1968. Computer simulation of the plastic deformation in face-centred cubic polycrystals and the rolling texture derived. physica status solidi (b) 25 (1), 337–344.

Leffers, T., 1979. A modified Sachs approach to the plastic deformation of polycrystals as a realistic alternative to the Taylor model. In: Strength of Metals and Alloys. Vol. 2. Pergamon Press, pp. 769–774.

Leffers, T., 1995. Long-range stresses associated with boundaries in deformed materials. physica status solidi (a) 149 (1), 69–84.

Lege, D., Barlat, F., Brem, J., 1989. Characterization and modelling of the mechanical behavior and formability of a 2008-T4 sheet sample. Int. J. Mech. Sci. 31 (7), 549–563.

Lipinski, P., Berveiller, M., 1989. Elastoplasticity of micro-inhomogeneous metals at large strains. Int. J. Plast. 5, 149–172.

Liu, Y., Castañeda, P. P., 2004a. Homogenization estimates for the average behavior and field fluctuations in cubic and hexagonal viscoplastic polycrystals. J. Mech. Phys. Solids 52 (5), 1175 – 1211.

Liu, Y., Castañeda, P. P., 2004b. Second-order theory for the effective behavior and field fluctuations in viscoplastic polycrystals. J. Mech. Phys. Solids 52 (2), 467 – 495.

Liu, Y., Gilormini, P., Castañeda, P. P., 2003. Variational self-consistent estimates for texture evolution in viscoplastic polycrystals. Acta Mater. 51 (18), 5425 – 5437.

Liu, Y., Gilormini, P., Castañeda, P. P., 2005. Homogenization estimates for texture evolution in halite. Tectonophysics 406 (3-4), 179 – 195.

Lloyd, S., 1982. Least squares quantization in PCM. Information Theory, IEEE Transactions on 28 (2), 129 – 137.

Ma, A., 2003. Implicit scheme of RVE calculation for fcc polycrystals. Comp. Mater. Sci. 27 (4), 471 – 479.

Masson, R., Bornert, M., Suquet, P., Zaoui, A., 2000. An affine formulation for the prediction of the effective properties of nonlinear composites and polycrystals. J. Mech. Phys. Solids 48 (6-7), 1203 – 1227.

Mathur, K., Dawson, P., 1989. On modeling the development of crystallographic texture in bulk forming processes. Int. J. Plast. 5, 67–94.

Mathur, K. K., Dawson, P. R., Kocks, U., 1990. On modeling anisotropy in deformation processes involving textured polycrystals with distorted grain shape. Mechanics of Materials 10 (3), 183 – 202.

Melchior, M. A., Delannay, L., 2006. A texture discretization technique adapted to polycrystalline aggregates with non-uniform grain size. Comp. Mater. Sci. 37 (4), 557 – 564.

Mercier, S., Molinari, A., 2009. Homogenization of elastic-viscoplastic heterogeneous materials: Self-consistent and Mori-Tanaka schemes. Int. J. Plast. 25 (6), 1024 – 1048.

Méric, L., Cailletaud, G., Gaspérini, M., 1994. F.e. calculations of copper bicrystal specimens submitted to tension-compression tests. Acta Metall. Mater. 42 (3), 921 – 935.

M'Guil, S., Ahzi, S., Youssef, H., Baniassadi, M., Gracio, J., 2009. A comparison of viscoplastic intermediate approaches for deformation texture evolution in face-centered cubic polycrystals. Acta Mater. 57 (8), 2496 – 2508.

M'Guil, S., Wen, W., Ahzi, S., 2010. Numerical study of deformation textures, yield locus, rolling components and Lankford coefficients for fcc polycrystals using the new polycrystalline ϕ-model. Int. J. Mech. Sci. 52 (10), 1313 – 1318.

M'Guil, S., Wen, W., Ahzi, S., Gracio, J., 2011. Modeling of large plastic deformation behavior and anisotropy evolution in cold rolled bcc steels using the viscoplastic ϕ-model-based grain-interaction. Mater. Sci. Eng. A 528 (18), 5840 – 5853.

Michel, J., Suquet, P., 2003. Nonuniform transformation field analysis. Int. J. Solids Struct. 40 (25), 6937 – 6955.

Miehe, C., Schröder, J., Schotte, J., 1999. Computational homogenization in finite plasticity. Simulation of texture development in polycrystalline materials. Comp. Meth. Appl. Mech. Engng. 171, 387–418.

Milton, G., 2002. The Theory of Composites. Cambridge Monographs on Applied and Computational Mathematics. Cambridge University Press.

Molinari, A., Canova, G., Ahzi, S., 1987. A self consistent approach of the large deformation polycrystal viscoplasticity. Acta metall. 35, 2983.

Morawiec, A., Oct 1998a. A note on mean orientation. J. Appl. Crystallogr. 31 (5), 818–819.

Morawiec, A., Jun 1998b. Comment on *Determination of a mean orientation from a cloud of orientations. Application to electron back-scattering pattern measurements* by Humbert *et al.* (1996). J. Appl. Crystallogr. 31 (3), 484.

Morawiec, A., 2004. Orientations and Rotations: Computations in Crystallographic Textures. Engineering Materials and Processes. Springer.

Moulinec, H., Suquet, P., 1998. A numerical method for computing the overall response of nonlinear composites with complex microstructure. Comp. Meth. Appl. Mech. Eng. 157 (1-2), 69 – 94.

Mura, T., 1987. Micromechanics of Defects in Solids. Kluwer Academic Publishers.

Musienko, A., Cailletaud, G., 2009. Simulation of inter- and transgranular crack propagation in polycrystalline aggregates due to stress corrosion cracking. Acta Mater. 57 (13), 3840 – 3855.

Nadeau, J. C., Ferrari, M., 2001. On optimal zeroth-order bounds with application to Hashin-Shtrikman bounds and anisotropy parameters. Int. J. Solids Struct. 38 (44-45), 7945–7965.

Nakamachi, E., Tam, N., Morimoto, H., 2007. Multi-scale finite element analyses of sheet metals by using SEM-EBSD measured crystallographic RVE models. Int. J. Plast. 23 (3), 450 – 489.

Nebozhyn, M., Gilormini, P., Castañeda, P., 2001. Variational self-consistent estimates for cubic viscoplastic polycrystals: the effect of grain anisotropy and shape. J. Mech. Phys. Solids 49, 313–340.

Neil, C., Wollmershauser, J., Clausen, B., Tomé, C., Agnew, S., 2010. Modeling lattice strain evolution at finite strains and experimental verification for copper and stainless steel using in situ neutron diffraction. Int. J. Plast. 26 (12), 1772 – 1791.

Nemat-Nasser, S., Hori, M., 1993. Micromechanics: Overall Properties of Heterogeneous Materials. Vol. 37 of North-Holland series in applied mathematics and mechanics. Elsevier Science Publishers B.V.

Nemat-Nasser, S., Obata, M., 1986. Rate-dependent, finite elasto-plastic deformation of polycrystals. Proc. R. Soc. London A 407 (1833), 343–375.

Nikolov, S., Lebensohn, R. A., Raabe, D., 2006. Self-consistent modeling of large plastic deformation, texture and morphology evolution in semi-crystalline polymers. J. Mech. Phys. Solids 54 (7), 1350 – 1375.

Novak, J., Kaczmarczyk, L., Grassl, P., Zeman, J., Pearce, C. J., 2012. A micromechanics-enhanced finite element formulation for modelling heterogeneous materials. Comp. Meth. Appl. Mech. Eng. 201-204, 53 – 64.

Ohser, J., Mücklich, F., 2000. Statistical Analysis of Microstructures in Materials Science. Wiley VCH.

Ostoja-Starzewski, M., 1993. Micromechanics as a basis of random elastic continuum approximations. Probabilistic Engineering Mechanics 8 (2), 107 – 114.

Ostoja-Starzewski, M., 2008. Microstructural Randomness and Scaling in Mechanics of Materials. Modern mechanics and mathematics. Chapman & Hall/CRC, Boca Raton, Fla.

Paufler, P., Schulze, G., 1978. Physikalische Grundlagen mechanischer Festkoerpereigenschaften. Vieweg, Braunschweig.

Pedersen, O. B., Leffers, T., 1987. Modelling of plastic heterogeneity in deformation of single-phase materials. In: Constitutive relations and their physical basis.

Pettermann, H. E., Huber, C. O., Luxner, M. H., Nogales, S., Böhm, H. J., 2010. An incremental Mori-Tanaka homogenization scheme for finite strain thermoelastoplasticity of mmcs. Materials 3 (1), 434–451.

Phan Van, T., Jöchen, K., Böhlke, T., 2012. Simulation of sheet metal forming incorporating EBSD data. J. of Mater. Process. Technol. 212, 2659–2668.

Ponte Castañeda, P., 1991. The effective mechanical properties of nonlinear isotropic composites. J. Mech. Phys. Solids 39 (1), 45–71.

Ponte Castañeda, P., 1992. New variational principles in plasticity and their application to composite materials. J. Mech. Phys. Solids 40 (8), 1757–1788.

Ponte Castañeda, P., 2002. Second-order homogenization estimates for nonlinear composites incorporating field fluctuations: I–theory. J. Mech. Phys. Solids 50 (4), 737 – 757.

Ponte Castañeda, P., Nebozhyn, M., 1997. Variational estimates of the self-consistent type for some model nonlinear polycrystals. Proc. R. Soc. London A 453, 2715–2724.

Ponte Castañeda, P., Suquet, P., 1998. Nonlinear composites. Adv. Appl. Mech. 34, 171–302.

Prakash, A., Lebensohn, R. A., 2009. Simulation of micromechanical behavior of polycrystals: finite elements versus fast fourier transforms. Modell. Simul. Mater. Sci. Eng. 17 (6), 064010 (16 pp).

Qiu, Y. P., Weng, G. J., 1992. A theory of plasticity for porous materials and particle-reinforced composites. J. Appl. Mech. 59, 261–268.

Qu, J., Cherkaoui, M., 2006. Fundamentals of Micromechanics of Solids. Wiley.

Quey, R., Dawson, P., Barbe, F., 2011. Large-scale 3D random polycrystals for the finite element method: Generation, meshing and remeshing. Comp. Meth. Appl. Mech. Eng. 200 (17-20), 1729–1745.

Raabe, D., Roters, F., 2004. Using texture components in crystal plasticity finite element simulations. Int. J. Plast 20, 339–361.

Ranganathan, S. I., Ostoja-Starzewski, M., 2008. Scaling function, anisotropy and the size of rve in elastic random polycrystals. J. Mech. Phys. Solids 56 (9), 2773 – 2791.

Renard, J., Marmonier, M., 1987. Étude de l'initiation de l'endommagement dans la matrice d'un matériau composite par une méthode d'homogénisation. Aerosp.Sci.Technol. 6, 37–51.

Reuss, A., 1929. Berechnung der Fließgrenze für Mischkristalle auf Grund der Plastizitätsbedingung für Einkristalle. Z. Angew. Math. Mech. 9, 49–58.

Roters, F., Eisenlohr, P., Hantcherli, L., Tjahjanto, D., Bieler, T., Raabe, D., 2010. Overview of constitutive laws, kinematics, homogenization and multiscale methods in crystal plasticity finite-element modeling: Theory, experiments, applications. Acta Mater. 58 (4), 1152 – 1211.

Rychlewski, J., Zhang, J., 1989. Anisotropy degree of elastic materials. Arch. Mech. 47 (5), 697–715.

Sachs, G., 1928. Zur Ableitung einer Fließbedingung. Z. Verein dt. Ing. 72, 734–736.

Salahouelhadj, A., Haddadi, H., 2010. Estimation of the size of the rve for isotropic copper polycrystals by using elastic-plastic finite element homogenisation. Comp. Mater. Sci. 48 (3), 447 – 455.

Sarma, G., Dawson, P., 1996. Texture predictions using a polycrystal plasticity model incorporating neighbor interactions. Int. J. Plast. 12 (8), 1023–1054.

Saylor, D., Fridy, J., El-Dasher, B., Jung, K.-Y., Rollett, A., 2004. Statistically representative three-dimensional microstructures based on orthogonal observation sections. Metall. Mater. Trans. A 35, 1969–1979.

Schaeben, H., 1997. A simple standard orientation density function: The hyperspherical de la vallée poussin kernel. physica status solidi (b) 200 (2), 367–376.

Schwartz, A., Kumar, M., Adams, B. (Eds.), 2000. Electron backscatter diffraction in materials science. Kluwer Academic.

Schwartz, A., Kumar, M., Adams, B., Field, D. (Eds.), 2009. Electron backscatter diffraction in materials science, 2nd Edition. Springer, New York, NY.

Segurado, J., Lebensohn, R. A., LLorca, J., Tomé, C. N., 2012. Multiscale modeling of plasticity based on embedding the viscoplastic self-consistent formulation in implicit finite elements. Int. J. Plast. 28 (1), 124 – 140.

Signorelli, J., Turner, P., Sordi, V., Ferrante, M., Vieira, E., Bolmaro, R., 2006. Computational modeling of texture and microstructure evolution in al alloys deformed by ECAE. Scripta Mater. 55 (12), 1099 – 1102.

Simmons, G., Wang, H., 1971. Single Crystal Elastic Constants and Calculated Aggregate Properties: A Handbook. The M.I.T. Press.

Simó, J., Hughes, T., 1998. Computational Inelasticity. Springer.

Sundararaghavan, V., Zabaras, N., 2006. Design of microstructure-sensitive properties in elasto-viscoplastic polycrystals using multi-scale homogenization. Int. J. Plast. 22 (10), 1799 – 1824.

Suquet, P., 1995. Overall properties of nonlinear composites: A modified secant moduli theory and its link with Ponte Castañeda's nonlinear variational procedure. C.R. Acad. des Sci. 320, Ser. IIb , 563–571.

Tadano, Y., Kuroda, M., Noguchi, H., 2012. Quantitative re-examination of taylor model for fcc polycrystals. Comp. Mater. Sci. 51 (1), 290 – 302.

Talbot, D., Willis, J., 1997. Bounds of third order for the overall response of nonlinear composites. J. Mech. Phys. Solids 45 (1), 87 – 111.

Talbot, D. R. S., Willis, J. R., 1985. Variational principles for inhomogeneous nonlinear media. IMA J. Appl. Math. 35, 39–54.

Talbot, D. R. S., Willis, J. R., 1994a. Upper and lower bounds for the overall properties of a nonlinear composite dielectric. I. Random microgeometry. Proc. R. Soc. London A 447 (1930), 365–384.

Talbot, D. R. S., Willis, J. R., 1994b. Upper and lower bounds for the overall properties of a nonlinear composite dielectric. II. Periodic microgeometry. Proc. R. Soc. London A 447 (1930), pp. 385–396.

Talbot, D. R. S., Willis, J. R., 1995. Upper and lower bounds for the overall properties of a nonlinear elastic composite. In: Inhomogeneity and Nonlinearity in Solid Mechanics. Proceedings, IUTAM Symposium on Anisotropy. Kluwer, Dordrecht.

Taylor, G., 1938. Plastic strain in metals. J. Inst. Metals 62, 307–324.

Tomé, C. N., Maudlin, P. J., Lebensohn, R. A., Kaschner, G. C., 2001. Mechanical response of zirconium-I. derivation of a polycrystal constitutive law and finite element analysis. Acta Mater. 49 (15), 3085 – 3096.

Torquato, S., 2002. Random Heterogeneous Materials: Microstructures and Macroscopic Properties. Springer.

Tóth, L. S., Van Houtte, P., 1992. Discretization techniques for orientation distribution functions. Textures and Microstructures 19 (4), 229–244.

Truesdell, C., Noll, W., 1965. The Non-Linear Field Theories of Mechanics. Vol. III/3 of Encyclopedia of Physics. Springer.

Van Houtte, P., 1982. On the equivalence of the relaxed Taylor theory and the Bishop-Hill theory for partially constrained plastic deformation of crystals. Mater. Sci. Eng. 55 (1), 69 – 77.

Van Houtte, P., Delannay, L., Samajdar, I., 1999. Quantitative prediction of cold rolling textures in low-carbon steel ba means of the LAMEL model. Texture Microstruct. 31, 109–149.

Van Houtte, P., Gawad, J., Eyckens, P., Bael, B. V., Samaey, G., Roose, D., 2012. Multi-scale modelling of the development of heterogeneous distributions of stress, strain, deformation texture and anisotropy in sheet metal forming. Procedia IUTAM 3, 67 – 75.

Van Houtte, P., Li, S., Seefeldt, M., Delannay, L., 2005. Deformation texture prediction: from the Taylor model to the advanced Lamel model. Int. J. Plast. 21 (3), 589 – 624.

Van Houtte, P., Rabet, L., 1997. Generalization of the relaxed constraints models for the prediction of deformation textures. Revue de métallurgie-cahiers d'informations techniques 94, 1483–1494.

Voigt, W., 1889. Über die Beziehung zwischen den beiden Elastizitätskonstanten isotroper Körper. Wied. Ann. 38, 573–587.

Walde, T., Riedel, H., 2007. Simulation of earing during deep drawing of magnesium alloy AZ31. Acta Mater. 55 (3), 867 – 874.

Walpole, L., 1966a. On bounds for the overall elastic moduli of inhomogeneous systems-i. J. Mech. Phys. Solids 14 (3), 151 – 162.

Walpole, L., 1966b. On bounds for the overall elastic moduli of inhomogeneous systems-ii. J. Mech. Phys. Solids 14 (5), 289 – 301.

Wenk, H.-R., 1999. A voyage through the deformed earth with the self-consistent model. Modell. Simul. Mater. Sci. Eng. 7 (5), 699–722.

Wierzbanowski, K., Jura, J., Haije, W. G., Helmholdt, R. B., 1992. F.C.C. rolling texture transitions in relation to constraint relaxation. Crystal Research and Technology 27 (4), 513–522.

Willis, J., 1977. Bounds and self-consistent estimates for the overall of anisotropic composites. J. Mech. Phys. Solids 25, 185–202.

Willis, J., 1981. Variational and related methods for the overall properties of composites. In: Yih, C.-S. (Ed.), Adv. Appl. Mech. Vol. 21. Elsevier, pp. 1 – 78.

Willis, J., 1983. The overall response of composite materials. ASME J. Appl. Mech. 50, 1202–1209.

Zeller, R., Dederichs, P. H., 1973. Elastic constants of polycrystals. Phys. Stat. Sol.(B) 55, 831–842.

Zhang, P., Karimpour, M., Balint, D., Lin, J., Farrugia, D., 2012. A controlled Poisson Voronoi tessellation for grain and cohesive boundary generation applied to crystal plasticity analysis. Comp. Mater. Sci. 64, 84 – 89.

Zhou, W., Wang, Z. (Eds.), 2007. Scanning Microscopy for Nanotechnology: Techniques and Applications. Springer.

**Schriftenreihe Kontinuumsmechanik im Maschinenbau
Karlsruher Institut für Technologie (KIT)
(ISSN 2192-693X)**

Herausgeber: Prof. Dr.-Ing. Thomas Böhlke

**Die Bände sind unter www.ksp.kit.edu als PDF frei verfügbar
oder als Druckausgabe bestellbar.**

Band 1   Felix Fritzen
**Microstructural modeling and computational homogeni-
zation of the physically linear and nonlinear constitutive
behavior of micro-heterogeneous materials.** 2011
ISBN 978-3-86644-699-1

Band 2   Rumena Tsotsova
Texturbasierte Modellierung anisotroper Fließpotentiale. 2012
ISBN 978-3-86644-764-6

Band 3   Johannes Wippler
**Micromechanical finite element simulations of crack
propagation in silicon nitride.** 2012
ISBN 978-3-86644-818-6

Band 4   Katja Jöchen
**Homogenization of the linear and non-linear mechanical
behavior of polycrystals.** 2013
ISBN 978-3-86644-971-8